Received
APR 21 2010
University Branch

NO LONGER PROPERTY OF
SEATTLE PUBLIC LIBRARY

The Kingdom Fungi

The Kingdom Fungi

The Biology of Mushrooms, Molds, and Lichens

Steven L. Stephenson

Timber Press
Portland | Cambridge

Copyright © 2010 by Steven L. Stephenson. All rights reserved.

Published in 2010 by Timber Press, Inc.

The Haseltine Building
133 S.W. Second Avenue, Suite 450
Portland, Oregon 97204-3527
www.timberpress.com

2 The Quadrant
135 Salusbury Road
London NW6 6RJ
www.timberpress.co.uk

Printing in China

Library of Congress Cataloging-in-Publication Data

Stephenson, Steven L.
  The kingdom fungi : the biology of mushrooms, molds, and lichens / Steven L. Stephenson. — 1st ed.
      p. cm.
  Includes bibliographical references and index.
  ISBN 978-0-88192-891-4
  1. Fungi. 2. Fungal molecular biology. 3. Fungi—Biotechnology. I. Title.
  QK603.S74 2010
  579.5—dc22
                                                                2009041293

A catalog record for this book is also available from the British Library.

To my wife, Barbara

Contents

Preface 9
Acknowledgments 11

  1 What Are Fungi? 13
  2 Fungi That Live in Water 26
  3 The Most Ubiquitous of All Fungi 46
  4 A Diversity of Form and Function 66
  5 Morels, Truffles, Cup Fungi, and Flask Fungi 85
  6 Mushrooms and Other Larger Fungi 100
  7 Lichens—More Than Just Fungi 128
  8 Slime Molds 146
  9 The Role of Fungi in Nature 163
10 Interactions of Fungi and Animals 186
11 Fungi and Humans 206
12 Fossil Fungi 227

Glossary 239
References 244
Index 252

Color photographs follow pages 48 and 160.

Preface

This book was written to provide a basis for understanding the diversity of what to the average person is a relatively little-known and underappreciated group of organisms—the fungi. Although it is difficult not to notice some fungi, including the molds that sometimes appear as fuzzy growths on spoiled food and the mushrooms that spring up in lawns immediately following a period of rainy weather during the summer or early fall, most people don't know much about these widely distributed, ecologically very important, and often exceedingly common organisms.

The overall purpose of the book is to introduce the reader to the biology, general structure, and morphological diversity of the "true" fungi as well as other funguslike (slime molds and water molds) and not-just-fungi (that is, lichens) organisms traditionally considered by the scientists (mycologists) who study fungi. Particular attention has been directed to examples of fungi that might be found in the home, encountered in nature, or mentioned in the media. The book should serve as a useful first text for the general naturalist, amateur mycologist, or interested layperson who simply wants to become more familiar with fungi.

The book is divided into 12 chapters. The first chapter provides a general introduction to the fungi as a group. In Chapters 2–8, the major taxonomic and ecological assemblages of fungi and funguslike organisms are discussed. Chapter 9 describes various roles that fungi have in nature, including their role as mycorrhizal associates of plants and their critical function in the decomposition of dead plant material. Chapter 10 discusses the various types of interactions that occur between fungi and animals, including the ants and termites that "cultivate" certain fungi in much the same manner as farmers cultivate crops. In Chapter 11, some of the more interesting or significant ways in which fungi have been involved in human affairs are described. This includes a consideration of fungi in the literature. Fungi have been around for a long time, and Chapter 12 provides information on what is known about fungi in the fossil record, including one seemingly improbable ancient fungus that was the largest living terrestrial organism 400 million years ago.

In preparing this book, I have made use of the information found in a number of textbooks on mycology. These include *The Biology of Fungi* (Ingold and Hudson 1993), *Introductory Mycology* (Alexopoulos et al. 1996), *The Fifth Kingdom* (Kendrick 2000), *Fungal Biology* (Deacon 2006), and *Introduction to Fungi* (Webster and Weber 2007). Additional information was derived from several field guides to the macrofungi. Among these are *Mushrooms of North America* (Miller 1973), *The Audubon Society Field Guide to North American Mushrooms* (Lincoff 1981), *Mushrooms of West Virginia and the Central Appalachians* (Roody 2003), and *North American Mushrooms* (Miller and Miller 2006). The first of these is particularly significant because it was used in my very first mycology course.

Acknowledgments

I am forever indebted to Orson K. Miller, Jr., for first introducing me to the fungi and then providing the encouragement that allowed me to pursue the mycological studies that ultimately shaped both my career and my life. The original suggestion to write a book on the natural history of fungi came from Dale Johnson, formerly at Timber Press. Although I was initially reluctant to take on such a project, Dale ultimately convinced me that it was something I should pursue. The fact that I had worked with Timber Press on two previous books was a major factor in my decision. I gratefully acknowledge the assistance of a number of people at the Press, without whom this project could not have been completed.

The Emily Johnson slide collection at the University of Arkansas represented the primary source of the images used in this book. Because Emily worked only with macrofungi during her career as a photographer, it was necessary for me to seek images of microfungi and certain other groups such as the lichens elsewhere. My quest for these additional images required considerable time and effort, but I ended up being quite satisfied with what I was able to obtain. I am enormously grateful to everyone who contributed images. Special thanks go to George Barron, Alan Bessette, Jerry Cooper, Jason Hollinger, Peter Letcher, Dan Mahoney, Joseph Morton, David Padgett, John Plischke, Bill Roody, Keith Seifert, Rod Seppelt, Scott Turner, and Ron Wolf, who went out of their way to provide multiple images. Other individuals who supplied images for the book include Harley Barnhart, Peter Bierdermann, Denise Binion, Lori Carris, Andrew Claridge, Ronnie Coffman, Randy Darrah, Kim Fleming, Charles Garratt, David Grimaldi, Camden Hackworth, Ian Hall, Michael Hood, Peter Johnston, Hans Kerp, Alena Kubátová, Gary Laursen, Estelle Levetin, Taylor Lockwood, Joyce Longcore, Robert Lücking, Daniel Mahr, Charles Meissner, Peter Nahum, Yuri Novozhilov, Mary Parrish, Tom

Pickering, Michael Pilkington, Dirk Redecker, Carlos Rojas, Alexander Schmidt, Martin Schnittler, John Shadwick, Cassius Stevani, Andy Swanson, Doug Waylett, Alex Weir, Merlin White, Ben Wrigley, and Wang Yun.

I would also like to express my appreciation to George Barron, Peter Bierdermann, Meredith Blackwell, Irwin Brodo, Harold Burdsall, Conrad Labandeira, Peter Letcher, Robert Lichtwardt, David Mitchell, Don Pfister, Bill Roody, Rod Seppelt, Keith Siefert, and Larissa Vasilyeva for reviewing various portions of the draft manuscript and making many valuable suggestions for its improvement.

Over my career, I have had an opportunity to interact with many individuals who have willingly shared their knowledge of fungi with me. I have been fortunate enough to have carried out field research with some of these individuals, and there is no better way to learn about fungi than directly in nature. My field research has taken me to many different regions of the world, and these opportunities to collect and study fungi were made possible because of financial support received from a variety of funding sources. The more important of these were the Australian Antarctic Division, the Australian Biological Resources Study (ABRS) Programme, the Fairmont State College Foundation, the National Geographic Society, the U.S. National Park Service, and the U.S. National Science Foundation.

Finally, no project of the magnitude required to produce a book could have taken place without the support and encouragement of my wife, Barbara. She has been my primary field assistant and constant companion on collecting trips since my days as a graduate student at Virginia Tech. Much of what I have managed to accomplish over my career could not have happened without her.

1
What Are Fungi?

Morphology of Fungi 14	Funguslike Organisms and Fungus Relatives 21
Growth of Fungi 15	Types of Fruiting Bodies 21
Reproduction in Fungi 17	Ecology and Spore Dispersal 22
Classification of Fungi 19	Numbers of Fungi 24
True Fungi 20	

Although the term "fungus" (plural: fungi) has at least some meaning to the average person, few people know very much about these widely distributed, ecologically very important, and often exceedingly common organisms. Fungi are abundant in nature, but they are often overlooked, usually underappreciated, and sometimes misunderstood. Their sudden appearance and disappearance, their frequent association with decaying organic matter, their vivid colors, fantastic shapes, and in some instances their poisonous properties, often cause fungi to be regarded as objects of mystery and sometimes even to be associated with the supernatural. Actually, fungi are among the most important inhabitants of the natural world, and there is little question that everyone should have a fundamental knowledge of what fungi are, what they look like, where they occur, and what they do.

The formal study of fungi is termed "mycology," and the scientists who consider at least some aspect of the biology or ecology of these organisms are known as mycologists. Mycology had its origin in botany, since fungi were once considered members of the plant kingdom. As a result, some of the terms that have been applied to the structures found in fungi are the same as those used for plants.

In some ways, fungi are indeed similar to plants, but unlike plants they lack the green pigment chlorophyll. Thus, fungi cannot produce their own food through photosynthesis. Instead, they obtain their food by breaking down dead organic matter or, in some cases, by attacking and living on, or within, living plants, animals, or even other fungi.

Fungi that depend upon dead organic matter as their food source are called saprotrophs, while those that feed on living hosts are called parasites if the host is harmed but not killed and pathogens if their presence produces a condition (called a disease) that has the potential of resulting in the death of the host. The distinction between parasite and pathogen is not necessarily absolute, and a parasite may become a pathogen over time or under a different set of circumstances. In addition, some fungi form a symbiotic relationship with the roots of trees and other plants. This relationship, which is called a mycorrhizal association, is mutually beneficial to both the plant and the fungus. The fungus enables the plant to take up nutrients that would otherwise be unavailable, and the plant provides nutrition for the fungus. The majority of plants, including such common forest trees as eucalypts, oaks, maples, hickories, and pines, are involved in these associations. In some instances, the mycorrhizal association is so essential to the plant that the latter would not survive without its fungal partner.

Fungi play a very important role in regulating natural processes. For example, it has been estimated that in a year several million leaves fall to the ground in each acre (0.4 ha) of a temperate deciduous forest. These leaves do not continue to pile up year after year because various saprotrophic fungi break them down. For that reason, essential nutrients in the leaves are recycled to the soil. Fungi are also the major group of organisms responsible for wood decay.

Although the vast majority of fungi are terrestrial, a number are aquatic forms, and some of these are of considerable ecological significance. For example, in small, well-aerated streams, certain aquatic fungi play a key role not only in the decomposition of organic matter introduced into streams but also as intermediates in food chains involving many aquatic insects and other invertebrates.

Morphology of Fungi

The vegetative body of all but the simplest fungi consists of a system of very finely branched, microscopic, threadlike structures called hyphae (singular: hypha). An entire mass of hyphae making up a particular fungus is known as a mycelium (plural: mycelia). The mycelium typically occurs in soil, leaf litter, or decaying wood, where the individual hyphae obtain the nutrients and water the fungus needs to grow (PLATE 1). After a period of growth and under favorable conditions of temperature and moisture, the mycelium usually produces one or more specialized fruiting structures, within or upon which the spores (literally the "seeds" of the fungus) are produced. The fruiting structure of a fungus is somewhat analogous to an apple on an apple tree, since it is the fruit of the mycelium. Most fruiting structures last for only a few days, but a mycelium may persist for a number of weeks, months, or even years.

As the fundamental structural unit of a fungus, a single hypha can be viewed as a thin, usually more or less transparent tube filled with cytoplasm and the various organelles typically associated with any eukaryotic cell. The tube itself represents the cell wall, and the latter is interrupted at places along its length by partitions, also referred to as cross walls or septa (singular: septum). In many fungi, septa occur at regular intervals, essentially delimiting individual cell-like compartments. In other fungi, septa are present only in special circumstances, such as to delimit a reproductive structure from an ordinary vegetative hypha. The term "septate" is used to describe fungi with regularly occurring septa, whereas those with few or no septa are referred to as coenocytic.

The hypha grows in length from the tip, and this growth can occur almost indefinitely if favorable conditions (a suitable supply of water and nutrients) persist. As it grows, the hypha spreads over or penetrates into a particular substrate, usually branching and thus expanding the mycelium of which it is part. More often than not, the mycelium occurs within the substrate and thus is not readily apparent in nature.

Growth of Fungi

The fact that some fungi appear to be capable of growing almost indefinitely under favorable conditions was borne out in an article that appeared in the April 2003 issue of the *Canadian Journal of Forest Research*. The authors of the article reported that a mycelium of *Armillaria ostoyae* (honey mushroom) apparently extended over a total area of 2200 acres (890 hectares) in the Malheur National Forest in the Strawberry Mountains of eastern Oregon (Ferguson et al. 2003). This "humongous fungus" was estimated to be more than 2000 years old and to have a total mass of as much as 605 tons (549 tonnes). If considered a single organism, this specimen would be the largest known organism in the world in terms of area and among the largest in the total amount of living biomass. Interestingly, one of the few organisms with a larger biomass is a clone of *Populus tremuloides* (quaking aspen) growing in the mountains of Utah that has been estimated to have a collective weight of more than 6000 tons (5443 tonnes). Since quaking aspen is an example of a tree that forms mycorrhizal associations with fungi, this clone could never have reached such a size without its fungus partner.

The most important component of the cell (or hyphal) wall of a fungus is a substance called chitin. Interestingly enough, chitin is also a major component of the exoskeleton of insects and other arthropods. However, the basic function of the fungal cell wall and the arthropod exoskeleton are quite different. The exoskeleton is a protective layer that also provides, as its name implies, an external structure to which the muscles of the organism in question are attached.

In fungi, the cell wall tends to be relatively thin, at least in most instances. While the thin cell wall does serve to maintain the form of the hypha, its main function is actually related to the manner in which fungi obtain their food. In striking contrast to the more familiar animals, which move from place to place to obtain food wherever it is available, and most plants, which produce their food through photosynthesis, most fungi (some simple motile forms represent exceptions) literally grow (generally through the elongation and proliferation of hyphae, as noted above) to reach a potential food source.

Upon reaching a food source, the fungus releases digestive enzymes (referred to as exoenzymes) into its immediate environment. These enzymes break down the large and relatively complex molecules of the food source (or substrate) into smaller molecules that can readily pass through the hyphal wall and then into the cytoplasm of the fungus. As such, digestion actually occurs on the outside of the fungus and not on the inside, as is the case for animals (including humans). Although the fungal method of obtaining food might seem less efficient than the animal method, the relative success of fungi suggests that this is not the case.

Reproduction in Fungi

Both sexual reproduction and asexual reproduction occur in fungi, although some species seem to have either lost the capability for sexual reproduction or do so only infrequently. As is true for all eukaryotic organisms, sexual reproduction involves the fusion of two nuclei and subsequent meiosis. Because this process allows for the occurrence of new combinations of genes in the offspring that are produced, it provides the basis for the changes needed to adapt to new and different environmental conditions. In short, it sets the stage for evolution. Asexual reproduction cannot do this. It simply involves the production of new individuals with the same combinations of genes. As such, the genetic "status quo" of the fungus is maintained. Although asexual reproduction might seem to offer few advantages, this is hardly the case. In fact, large numbers of asexual spores (and thus potential new individuals) can be produced in a short period of time, which greatly enhances the rapid colonization of new food sources as they become available.

Interestingly, it is now known that genetic recombination without meiosis can take place in certain fungi that only reproduce asexually. This special type of genetic recombination (referred to as parasexual reproduction) involves the fusion of two genetically different haploid nuclei to form a diploid nucleus, which then undergoes a loss of chromosomes during successive mitotic divisions until it is once again haploid but with a different complement of chromosomes than either of the original nuclei. Parasexual reproduction is thought to be comparatively rare in nature, but it does allow some degree of genetic recombination to occur in the absence of meiosis.

Although the majority of fungi appear to be capable of reproducing both sexually and asexually, the structures involved in each of the two types of reproduction are not typically produced at the same time. Moreover, a major difference often exists with respect to which is more common and/or important for a particular type of fungus. In general, asexual reproduction tends to be predominant in those fungi considered more primitive.

The reproductive structures produced by fungi are exceedingly important for another reason. With very rare exceptions, it is difficult or even impossible to identify a fungus from the features that can be observed for vegetative hyphae. Instead, our concepts of what constitutes a particular kind of fungus are based almost entirely upon the morphology of the reproductive structures. Here it should be pointed out that the application of the techniques of modern molecular biology to the study of fungi has provided a body of new data that has forced mycologists to reconsider some of these concepts, especially in terms of degrees of relatedness among different fungi.

One problem that exists, especially for many of the microscopic fungi to be considered in Chapter 3, is that what appear to be two closely related fungi with morphologically similar (sometimes essentially identical) asexual reproductive structures produce totally different sexual reproductive structures. The reverse situation (similar sexual reproductive structures but totally different asexual reproductive structures) also occurs. Prior to the use of the molecular techniques mentioned above, this situation posed a major problem for mycologists. The availability of the body of new data from molecular-based studies has given mycologists a new perspective on relationships among these fungi.

In many fungi, the reproductive structure produced is relatively simple and often microscopic, but in others it is large and complex. These larger fruiting structures are usually referred to as fruiting bodies (or sometimes fruit bodies). The various cultivated mushrooms, grown commercially on a large scale throughout much of the world and available for sale in most local supermarkets, represent examples of such structures that are familiar to almost everyone.

Fruiting bodies vary considerably in size, shape, color, and the circumstances under which they are found. Some come in rather bizarre shapes (PLATE 2), while others are objects of considerable beauty. Just where a fruiting body is found usually reflects the location of the mycelium from which it was produced. Accordingly, fruiting bodies that occur on wood are most likely to be associated with the mycelium of a wood-decomposing fungus, whereas those found on the ground are probably produced by fungi that either decompose dead leaves and other types of plant debris or form mycorrhizal relationships with nearby trees. Although the fruiting structures produced by microscopic fungi, a group that actually includes the vast majority of fungi, are relatively simple, many of them are quite intricate, sometimes to the point of defying the imagination (PLATE 3).

Classification of Fungi

As is the case for all organisms, for a particular kind of fungus to be formally recognized by mycologists, it must have been given a taxonomic name in accordance with certain internationally accepted rules. This name is a binomial that consists of a unique combination of a generic name (genus) followed by a specific epithet (species). For example, *Agaricus* is the genus name and *bisporus* the specific epithet of the most common cultivated mushroom. The binomial that has been given to a particular species is complete only when it is followed by the authority name, the last name (or an abbreviation of the name) of the person (or persons) who first described the species in question.

Since taxonomic concepts are subject to change, a species can sometimes be transferred to a genus other than the one to which it was assigned when originally described. When this happens, the name of the person who formally proposed the change is added to the authority, and the original name is placed in parentheses. For example, *Agaricus bisporus* (J. E. Lange) Pilát indicates that the Danish mycologist Jakob Lange was the first to apply a name to this fungus, but he considered it to be a member of the genus *Psalliota*, which is no longer recognized. Later, the Czech mycologist Albert Pilát placed the fungus in the genus *Agaricus*, which is the concept still recognized by most modern mycologists.

Genus and species are but two levels in a hierarchical system of classification in which organisms are grouped on the basis of common characteristics. Two or more species that share many of the same characteristics are placed in the same genus (plural: genera), while genera with common characteristics are grouped in the same family. For example, *Agaricus bisporus* and the closely related and morphologically very similar *A. campestris* (meadow mushroom) are both members of the genus *Agaricus* (PLATE 4). The genus belongs to the family Agaricaceae, which includes such other genera as *Chlorophyllum*, *Lepiota*, and *Leucoagaricus*. Likewise, families with common characteristics are grouped in the same order. Correspondingly, orders that share the same characteristics are grouped in the same class, classes with common characteristics in the same phylum (plural: phyla), and phyla in the same kingdom, in this case the kingdom Fungi.

On some occasions, a subdivision of one of the seven main levels of classification is used. For example, three subphyla have been recognized for the phylum Basidiomycota. One of these, the subphylum Agaricomycotina, includes most of the fungi that people usually consider as mushrooms. As such, *Agaricus bisporus* belongs to this subphylum.

The term "taxon" (plural: taxa) can be used in a general sense to refer to any level of classification. *Agaricus bisporus* and the subphylum to which it belongs are two different taxa, and the same would be true for any two of the genera mentioned above (for example, *Chlorophyllum* and *Lepiota*). Conversely, all true fungi, since they are all placed in the same kingdom, make up a single taxon.

True Fungi

In most treatments of the kingdom Fungi, four phyla have been accepted as being "true" fungi. These are the Chytridiomycota, Zygomycota, Ascomycota, and Basidiomycota. However, research into fungal phylogenetics carried out by numerous mycologists in the context of two projects ("Deep Hypha" and "Assembling the Fungal Tree of Life") funded by the National Science Foundation has provided additional insight into the possible evolutionary relationships of the various members of these four phyla (Blackwell et al. 2006). In brief, some of the organisms traditionally assigned to two of the phyla (Chytridiomycota and Zygomycota) have been removed and placed elsewhere, sometimes in their own phylum. It now appears appropriate to recognize the phylum Glomeromycota for a group of fungi previously thought to be members of the Zygomycota and the phylum Blastocladiomycota for some of the fungi once considered to belong to the Chytridiomycota. In addition, numerous classes and orders within the Chytridiomycota, Zygomycota, Ascomycota, and Basidiomycota have been redefined. To the extent possible, the taxonomic treatment used herein reflects these recent changes.

Note should be made of the fact that the groups of fungi recognized as phyla in the system of classification outlined above were once considered as taxonomic classes. For example, the fungi now placed in the Ascomycota made up the class Ascomycetes, while members of the Basidiomycota were assigned to the class Basidiomycetes. These terms are still used in an informal sense (that is, "ascomycetes" and "basidiomycetes"), especially by amateur mycologists but also by some (especially older) professional mycologists. Oftentimes, the terms are further abbreviated to "ascos" and "basidos," which are not officially sanctioned technical names. Nevertheless, they are convenient to use when referring to a particular example from one of the two phyla.

Funguslike Organisms and Fungus Relatives

In addition to the "true" fungi, several other groups of organisms have long been studied almost exclusively by mycologists, since they are "funguslike" in some respects. The best example is the group known as water molds, which have often been treated in the context of the true fungi as the phylum Oomycota. These organisms have a vegetative body consisting of hyphal-like filaments that superficially resemble the hyphae of fungi, and they obtain their food in the same manner. Water molds also possess a number of other features that indicate they are not closely related to fungi. A second example is represented by the slime molds, some of which produce fruiting bodies similar to, albeit usually much smaller than, those of certain macrofungi. Other than this, slime molds share few other features in common with the true fungi, although traditionally studied along with fungi and included in most textbooks of mycology.

The members of yet another group (the lichens) are more than just fungi. These organisms, which will be considered in detail in Chapter 7, are fungi that have established an intimate biological relationship with another organism (an alga) that enables them to survive under conditions that could not be tolerated by the fungus alone. This "composite" organism is usually very different in appearance from what it would be with only the fungus present, and the fungal component might not be recognized as such. Because mycologists have traditionally grouped all of these organisms with the fungi, they are considered herein.

Types of Fruiting Bodies

Most of the large, conspicuous fruiting bodies encountered in nature are produced by members of the Basidiomycota. The various different kinds of fungi that make up this phylum are distinguished from one another by the location of the spore-producing hyphae and by the overall shape of the fruiting body. Among the more familiar members of the Basidiomycota are the mushrooms, polypores, puffballs, boletes, and chanterelles.

A few fungi with conspicuous fruiting bodies (for example, morels and truffles) belong to the Ascomycota. Many of the larger members of this phylum have a cup- or bowl-shaped fruiting body with the spore-producing hyphae forming a layer over the upper surface. A few examples produce fruiting bodies with shapes more difficult to characterize. Such is the case for the morels, which are among the most highly prized edible fungi.

In contrast to the Basidiomycota and Ascomycota, members of the Chytridiomycota and Zygomycota are generally too small to be easily seen in nature, although it is certainly possible to notice the presence of their mycelia on certain substrates. The fuzzy gray growth that sometimes forms on a piece of bread is often the mycelium of a member of the Zygomycota.

Ecology and Spore Dispersal

As a group, fungi utilize an incredible diversity of organic substrates as sources of food and occur in many different situations. Indeed, it is difficult to think of any substance that they are unable to exploit or any habitat where they are completely absent. The vast majority of fungi are terrestrial, but numerous species can be found in fresh water and most members of the Chytridiomycota are found in such habitats. Fungi are not well represented in the marine environment, although they can be common in the intertidal zone. Surveys of fungi carried out in polar regions have yielded surprising numbers of species. Some of these "cryophilic" (low-temperature-loving) fungi are so well adapted to the low temperatures under which they survive that they do not grow under more moderate temperatures. On the other end of the temperature spectrum, certain species of fungi are exceedingly resistant to high temperatures and occur in hot springs and deserts.

The spores produced by fungi tend to be relatively small, with those of many species no more than about 10 μm in diameter (a micrometer, μm, is equal to one-thousandth of a millimeter, or 0.0004 inch). Any object this size, if it gets into the air, has the potential of being carried some distance by the wind. This trait would suggest that long-distance dispersal is possible for fungi and that a particular species might be expected to occur over large areas of the world. However, what is the evidence that such long-distance dispersal actually takes place?

It has long been recognized that various small particles, including dust, spores, and bacteria, can be carried long distances by wind. The British mycologist Miles Joseph Berkeley (1857) concluded: "The trade winds, for instance, carry spores of fungi mixed with their dust, which may have traveled thousands of miles before they are deposited." In the mid-1930s, the famous aviator Charles Lindbergh helped carry out an experiment that involved exposing sterile, oil-coated microscope slides directly to the air by way of a long metal arm extending from an airplane flying over the North Atlantic (Meier and Lindbergh 1935). An examination of these slides in the laboratory revealed a variety of fungal spores along with pollen grains, algae, and diatoms.

We now know that enormous numbers of fungal spores and other types of organic particulate matter (especially pollen grains at certain times of the year) are suspended in the atmosphere, as anyone who suffers from "hay fever" is very much aware. Aerobiologists such as Estelle Levetin at the University of Tulsa use various filtering devices to concentrate airborne pollen grains and spores, and the samples obtained provide some idea of the diversity of fungal spores actually present in the atmosphere (PLATE 5).

A process that could transfer enormous numbers of spores and microorganisms into the atmosphere was identified in the late 1990s, when satellite images revealed the astonishing magnitude by which desert soils are aerosolized into giant clouds of dust in northern Africa (Griffin et al. 2001). These clouds of dust frequently move across the Atlantic Ocean and reach the Caribbean, Central America, northern South America, and the southeastern United States, where the particles they contain (including fungal spores) are deposited. Similar long-range movements of dust have been demonstrated for other parts of the world, including from Asia across the Pacific Ocean to western North America, and from Australia to New Zealand. One particularly large dust cloud originating in China actually moved eastward all the way across the Pacific, North America, and the Atlantic to reach Europe, thus traveling most of the way around the world.

Clearly, airborne fungal spores would have the potential of being dispersed by wind over considerable distances, and some of the more commonly encountered species of microfungi do appear to be ubiquitous, since they turn up anywhere in the world if the appropriate sampling is carried out. Many other species of fungi (especially for certain groups such as many macrofungi) appear to be restricted to particular geographic regions of the world. For example, the assemblage of macrofungi found in Europe is clearly distinct from the assemblage found in North America, although there are species that occur in both places. In many instances, the presence of a given species in one region but not another can be attributed to the fact that the distribution of the fungus in question reflects some type of relationship (for example, a mycorrhizal association) with a type of tree that is confined to a single region. In other instances, just why a fungus has a more limited distribution than would have been anticipated if long-distance dispersal is indeed a common phenomenon is a lot less obvious.

Numbers of Fungi

The total number of different species of fungi found on the earth is not yet known, but it almost certainly exceeds one million and some estimates are appreciably higher (Hawksworth 2001). Estimates of fungal biodiversity are based on extrapolations from data on the numbers of species of both vascular plants and fungi known from a particular region of the world (for example, Great Britain and Ireland) for which both groups of organisms were considered very well known. Numbers of fungi always exceed numbers of vascular plants, but a ratio can be generated and then applied to the entire world. For example, the ratio of fungi to plants in Great Britain and Ireland is approximately six to one. Since the total number of vascular plants worldwide is thought to be about 270,000, applying this ratio yields a probably conservative estimate of 1.5 million fungi.

It is possible that the six-to-one ratio seriously underestimates the number of fungi in some parts of the world. An intensive survey of the fungi associated with two species of palms in Southeast Asia suggested a fungus-to-plant ratio of thirty-three to one (Fröhlich and Hyde 1999). Since no more than about 100,000 species of fungi have been described to date, a rather noticeable discrepancy is apparent when this number is compared with any of the estimates that have been generated.

Clearly, there are tremendous numbers of fungi yet to be discovered. Mycologists agree that this is indeed the case, and it is the result of several factors. Foremost among these is the simple fact that there are large areas of the world, particular habitats and certain examples of substrates in those habitats, in which the fungi are woefully understudied. Many regions of the tropics and subtropics are good examples. On average, more than 1000 new species of fungi are described by the mycologists of the world each year, and perhaps half of these are fungi from the tropics and subtropics. Even if the rate at which new species are described is increased several fold, it is exceedingly doubtful that the majority of fungi with which humans share this planet will ever have formal names. However, a good case could be made for considering this to be irrelevant. The only thing that seems necessary is an awareness of the tremendous diversity that exists for fungi.

2
Fungi That Live in Water

Phylum Chytridiomycota 27	Late Blight of Potato 36
Chytrid Reproduction 28	Grapevine Downy Mildew 37
Chytrids in Nature 29	Dieback 38
Chytrid Disease in Amphibians 29	**Class Trichomycetes** 40
Phylum Blastocladiomycota 31	**Aquatic Members of Phylum Ascomycota** 42
Water Molds 33	**Aquatic Hyphomycetes** 42
Asexual Life Cycle in Water Molds 34	Stream Ecology and Food Chains 43
Sexual Reproduction in Water Molds 34	Aero-aquatic Fungi 44
Water Molds as Parasites 36	Aquatic Macrofungi 44

When I first studied fungi in a botany class as an undergraduate more than 40 years ago, the classification given in the textbook used for the class was not quite the same as the one used today. The biggest difference is that all of the more primitive fungi, many of which are found in water, were grouped in the class Phycomycetes. The term literally translates as "algal-like fungi" and reflects the fact that many members of this assemblage are found in water or at least in damp places. However, as noted in Chapter 1, the water molds, one of the three main assemblages of organisms once thought to comprise the Phycomycetes, are not fungi at all, although they are similar to fungi in a number of ways. Their vegetative body consists of a system of microscopic, thread-like filaments, and they obtain food in the same manner as fungi (that is, by producing exoenzymes that break down organic materials into smaller molecules that can be absorbed).

In short, these organisms look like fungi and seem to act like fungi, but there are a number of noteworthy differences between water molds and fungi. First, all "true" fungi have cell walls composed of chitin, but the cell walls of water molds consist of cellulose-like compounds similar to those found in the cell walls of plants. Moreover, the hyphalike filaments produced by water molds are larger than the hyphae of most true fungi, and there are no septa except to delimit reproductive structures. Only the members of the Blastocladiomycota, Chytridiomycota, and Zygomycota share the latter feature. Second, the nuclei found within the filaments that make up the vegetative body of a water mold are diploid and not haploid as is the case in almost all fungi. Nevertheless, water molds are almost invariably studied by mycologists and will be considered in this chapter.

As they were circumscribed at the time I first encountered the group as an undergraduate, the Phycomycetes also included, in addition to the water molds, the three groups of true fungi (Blastocladiomycota, Chytridiomycota, and Zygomycota) mentioned above. At that time, the Blastocladiomycota was not yet considered distinct from the Chytridiomycota. All three of these phyla are characterized by vegetative hyphae that lack septa, but the members of the Zygomycota are terrestrial and will be considered in Chapter 3.

Phylum Chytridiomycota

The members of the Chytridiomycota, known as chytrids or chytrid fungi, are thought to be the most primitive of the true fungi. There are approximately 1000 described species of chytrids, and members of the group are essentially ubiquitous, being found in most types of habitats and occurring from the polar regions to the tropics. The majority of species are thought to occur in terrestrial habitats such as forest, agricultural, and desert soils, but the group is also well represented in freshwater habitats, including streams, ponds, lakes, marshes, and bogs. A few species tolerate salt water and can be found in estuaries.

Most chytrids are saprotrophs, feeding upon plant and animal debris introduced into the habitats in which they occur. Some species are parasites of algae or small aquatic animals (PLATE 6), particularly those chytrids found in aquatic habitats. A few other chytrids are parasites of vascular plants and are directly important to humans because some of the plants involved are agricultural crops, including corn, potatoes, cabbage, and alfalfa.

Recently, much to the surprise of mycologists, certain organisms, previously thought to be protozoans and found in the rumen (the specialized region of the gut where microbial processing of plant material takes place) of animals such as cattle and sheep, were actually very specialized chytrids. Obviously, a rumen is a very different kind of aquatic habitat. Originally placed in the Chytridiomycota, these chytridlike forms are now considered to constitute a separate phylum, the Neocallimastigomycota.

Unlike most other fungi, chytrids form no true mycelium. Most species are best regarded as unicellular, with the entire vegetative plant body (or thallus) consisting of a single cell-like unit. The thallus may produce a number of slender, somewhat rootlike structures (called rhizoids) that penetrate the plant or animal debris upon which the chytrid is feeding.

Chytrid Reproduction

Chytrids are the only true fungi that reproduce by motile spores (called zoospores), typically propelled by a single, posteriorly directed flagellum. The word *chytrid* is derived from the Greek *chytridion* (meaning "little pot"), which refers to the zoosporangium, the structure within which the zoospores are produced.

Usually, the thallus of a fully mature chytrid consists of just the zoosporangium, which is often more or less spherical, and a system of rhizoids. The zoosporangium is derived from the body of the original zoospore that came to rest on a suitable substrate, formed a thin cell wall around itself, and put out a system of rhizoids to penetrate the substrate.

In addition to digesting and absorbing food, the rhizoids also serve to anchor the chytrid to the substrate. Growth is limited in chytrids, and by the time the food represented by the substrate upon which it occurs has been utilized, zoospores are delimited within the zoosporangium. These zoospores eventually emerge through specialized exit openings that develop in the wall of the zoosporangium (PLATE 7). In some chytrids, a well-defined circular cap (referred to as an operculum) develops at the discharge point. Once released from the zoosporangium, the zoospores literally swim off in search of new substrates to exploit.

There is considerable variation among chytrids in the structure of the thallus, where the thallus occurs in relation to the substrate, and the morphology of the reproductive structures. In some examples, the thallus is produced on the surface of the substrate upon which the zoospore comes to rest, a type of situation referred to as epibiotic. Other chytrids are endobiotic, which simply means that they live entirely within their hosts.

Sexual reproduction is the exception rather than the rule in chytrids. When sexual reproduction does occur, it may involve two zoospores, each derived from a different thallus, essentially functioning as gametes and fusing together to form a zygote, which actually remains motile in some species. In most instances, the zygote ultimately gives rise to a thick-walled resting stage, within which meiosis occurs. Mitosis and the production of zoospores follow, but usually not immediately, and this resting stage in the life cycle is capable of surviving when environmental conditions are unfavorable, such as during winter. The production of a resting stage is common among chytrids but there is little evidence that it is always the result of sexual reproduction.

Chytrids in Nature

Since chytrids are so extremely small, most species are recognizable only under the high power objective of a microscope. The usual way to obtain chytrids for study is to "bait" a sample of water or a soil suspension for those species that occur in these two habitats. Among the baits commonly used are small pieces of cellophane, insect wings, and pollen grains (especially from pine). Chytrids often appear on such substrates within a few days, and their frequency of occurrence and sheer numbers give some indication of just how common and widespread these organisms are in nature (PLATE 8).

Chytrid Disease in Amphibians

The vast majority of chytrids are innocuous, with seemingly little reason for the average person to even know that they exist. One exception has attracted a great deal of attention in the public media. Since about 1980, dramatic declines in populations of frogs and other amphibians have been reported from all over the world. During this period of time, more than 100 species of amphibians may have become extinct, and perhaps half of the 6000 extant species appear to be threatened.

One of the first amphibians to disappear was the golden toad (*Bufo periglenes*). The golden toad was restricted to a small area of forest in the mountains of Costa Rica. Because of its distinctive color, the toad was often featured prominently in magazine articles and other printed literature relating to the tropical cloud forests of the Monteverde region of Costa Rica. Moreover, since the species had been the subject of an ongoing scientific investigation, its distribution and ecology were relatively well known. Populations of the toad suddenly crashed in 1987, and the species had disappeared completely by 1989. Other species at Monteverde, including the Monteverde harlequin frog (*Atelopus varius*), also disappeared at the same time. Because these species occurred in an essentially pristine and little-disturbed area of forest in the Monteverde Cloud Forest Reserve, it seemed unlikely that these extinctions could be related to local human activities, as has been the case for certain other amphibians in other areas of the world. As such, these extinction events were of particular concern to biologists.

In 1998, following investigations of the possible cause for large-scale frog deaths in Australia and Panama, research teams in both places concluded that a previously unknown and thus undescribed species of pathogenic fungus was partially if not largely responsible for the observed mortality. The most important clues were the spherical bodies consistently present inside skin cells of the dead frogs (PLATE 9). Joyce Longcore, a mycologist at the University of Maine, had isolated the same fungus in 1997 from a captive blue poison dart frog (*Dendrobates azureus*) that had died at the National Zoological Park in Washington, D.C. Longcore, who happened to be an expert on chytrids, immediately recognized the organism as a member of this group of fungi. This discovery was surprising, since chytrids are not generally pathogenic and they were not previously known to be capable of causing disease and death in any vertebrate animal. In 1999, Longacre and two of her colleagues formally described this deadly chytrid as *Batrachochytrium dendrobatidis* (Longcore et al. 1999).

The disease caused by *Batrachochytrium dendrobatidis* is called chytridiomycosis. Amphibians infected by the disease generally develop pronounced skin lesions, and it is thought that death occurs because the presence of the fungus makes it impossible for normal respiration to occur through the skin. The time from infection to death has been found to be as little as one to two weeks in experimental animals.

Studies have established that the fungus has been present in Australia since at least 1978 and was already present in North America in the 1970s. The first known occurrence of the chytrid infection was in the African clawed frog (*Xenopus laevis*). Because specimens of *Xenopus* are sold in pet stores and used in laboratories around the world, it is possible that the chytrid fungus may have been exported from Africa.

A number of perplexing questions about *Batrachochytrium dendrobatidis* still remain. One of the more important of these has to do with the degree to which the fungus affects the host. Longcore, who examined individual frogs from populations throughout Maine, discovered that *B. dendrobatidis* was common and widespread, but appeared to cause only a mild, sublethal infection and not the nearly 100 percent mortality reported from other regions of the world.

As a group, amphibians are characterized by a life cycle that consists of an aquatic stage (where larval amphibians such as the tadpoles of frogs occur) and a terrestrial stage (where the adults usually occur). It is in the aquatic stage that they are potentially exposed to *Batrachochytrium dendrobatidis*. Moreover, because their skins are highly permeable, amphibians are probably more susceptible to infection than other vertebrates such as certain birds or mammals that live in aquatic habitats. Without question, human activities are responsible for losses in global biodiversity, but amphibians appear to be suffering more than most other organisms. *Batrachochytrium dendrobatidis*, a member of an otherwise relatively insignificant group of fungi, certainly adds to their problems.

Phylum Blastocladiomycota

As noted above, fungi now assigned to the Blastocladiomycota were considered as belonging to the Chytridiomycota until molecular data and the results obtained from detailed examination of their cell ultrastructure indicated that they actually constituted a distinct group. These fungi are mostly saprotrophs that occur in soil, water, mud, or various types of organic debris found in aquatic habitats. A few species are pathogens of plants, animals, or fungi.

There is a considerable range of morphological variation in the Blastocladiomycota. In relatively simple forms, the thallus consists of little more than a single cell-like unit that is similar in appearance to the condition that exists for many chytrids. In other members of the phylum, the thallus is more complex, often consisting of a group of well-developed rhizoids, a trunklike portion, and numerous branched hyphae that arise from the latter. The reproductive structures are formed at the tips of the branches.

The best-known member of this group of fungi is *Allomyces*. Species of *Allomyces* are saprotrophs found in mud or soil in temperate to tropical regions. One unusual feature of *Allomyces* and some other taxa in the Blastocladiomycota is the alternation of distinct haploid (gametothallus) and diploid (sporothallus) generations in the life cycle. The thallus produced in each of these generations is morphologically identical except for the types of reproductive structures produced. The haploid gametothallus produces a smaller "male" gamete and a larger "female" gamete in specialized gametangia, while the diploid sporothallus produces two types of sporangia—mitosporangia and meiosporangia. The mitosporangium is a thin-walled, ephemerical structure that gives rise to a diploid, uniflagellate zoospore that can establish new sporothalli, while the meiosporangia are thick-walled and resistant enough to survive for a couple of decades.

The meiosporangium, also sometimes called a resting sporangium, is the overwintering stage in the life cycle in *Allomyces*. Eventually, meiosis occurs within a meiosporangium, the outer wall of the latter breaks open, and haploid, uniflagellate meiospores are liberated. These develop into new gametothalli.

This fungus is relatively easy to obtain in laboratory culture by placing dried samples of soil in a container of water and then adding an appropriate bait. Surprisingly, split dry hemp seeds are among the best seeds to use. Hemp is simply another name for the plant known as marijuana (*Cannabis sativa*), and its seeds are (for obvious reasons) not readily available. It is sometimes possible to purchase seeds that have been rendered sterile from some biological supply companies, and this is the source used by mycologists. Because *Allomyces* is easily isolated from nature and then maintained in culture, it has received considerable study by mycologists.

The genus *Blastocladiella* consists of about a dozen species in which the thallus is comparatively simple and thus more like that of a chytrid. In most cases, it consists of a short unbranched axis bearing rhizoids at one end and a single, cell-like sporangium at the other. These fungi are commonly found in soil and water. *Blastocladiella emersonii* is the best-known member of the genus and has been studied extensively in the laboratory.

Water Molds

Although water molds are not true fungi, they are morphologically similar to fungi and occur in some of the same habitats. Members of the group are common freshwater organisms and also occur in moist soil. Most of the approximately 600 species are saprotrophs, living on animal and plant debris. A few are parasites of no great significance on algae and other forms of aquatic life. Some of the others are among the most destructive parasites known to mankind. As will be discussed later in this chapter, several have had a profound effect upon the course of human history.

Like the chytrids, water molds produce motile zoospores. These zoospores differ from those of chytrids by being biflagellate, with one flagellum projecting forward during swimming and the other trailing behind. Interestingly, the two flagella are of different types, with the anteriorly directed flagellum bearing numerous fine lateral filaments along its main axis (a condition that accounts for the name tinsel flagellum) and the posteriorly directed having a smooth axis (thus the name whiplash flagellum). Moreover, as will be described below, water mold produce two distinctly different types of zoospores.

Water molds are unlike chytrids in another important respect, because they do produce a mycelium. Often, a water mold mycelium can be observed directly in nature as a white, cottony halo that forms around the body of a dead insect or fish floating on the surface of an aquarium or a quiet pool. In fact, mycologists usually obtain specimens of water molds in the same manner as described earlier for chytrids, although the actual baits used are somewhat different. Dead insects, split dry hemp seeds, or sesame seeds are among the more commonly used baits for water molds.

Asexual Life Cycle in Water Molds

The zoospores of water molds are produced within a zoosporangium, which develops as a somewhat expanded, cigar-shaped tip of a vegetative hypha (PLATE 10). As it forms, the contents of the zoosporangium are delimited from the subtending hypha by a septum. Typically, numerous zoosporangia are produced by a single mycelium. At first the zoosporangium contains a multinucleate mass of protoplasm, but eventually this mass separates into numerous uninucleate zoospores. When mature, these zoospores can be observed moving about within the zoosporangium. At this point, a thin and slightly protruding area at the tip of the zoosporangium breaks down, and the zoospores are released. Each zoospore is pear-shaped, and the two flagella arise from the shorter end.

The discharged zoospores swim actively about for a short time but eventually they stop, withdraw their flagella, and become enclosed by a thin wall (or cyst). This period of encystment lasts for a few hours, after which the zoospore reemerges. Interestingly, the basic form of the zoospore undergoes a change during this process. When released from the cyst, it is more or less bean-shaped, with the two flagella arising laterally. As such, it is possible to distinguish two types of zoospores, the form produced initially (the primary zoospore) and the one derived from it (the secondary zoospore). The secondary zoospore also has a period of encystment, which may occur more than once. Presumably, the formation of the resistant cysts by the zoospores of water molds is a mechanism for survival under less than optimal conditions. Eventually, the cyst germinates in much the same way as would be expected of a fungal spore, giving rise to the first hypha of a new vegetative mycelium.

The asexual life cycle outlined above is characteristic of members of the genus *Saprolegnia* and some of its relatives. *Saprolegnia* is an exceedingly common water mold that is easily isolated from samples of fresh water and soil. Because of this fact, it often serves as the example of a water mold studied in the laboratory.

Sexual Reproduction in Water Molds

Achlya is another common and widespread representative of the same taxonomic order (the Saprolegniales) to which *Saprolegnia* belongs.

Members of the two genera are similar morphologically, but they differ in a couple of important respects, two of which relate to the zoosporangium and zoospores. First, the primary zoospores of *Achlya* commonly encyst almost immediately upon being discharged, forming a cluster of cysts at the tip of the zoosporangium. When produced, the secondary zoospores emerge from this cluster. Second, in *Saprolegnia* the septum at the base of the empty zoosporangium bulges to produce a new hyphal tip that grows forward and develops into a second zoosporangium located inside of the original one (PLATE 11). In *Achlya*, this process does not happen. Instead, a new zoosporangium develops alongside the original one.

When conditions are favorable for sexual reproduction to occur, vegetative hyphae give rise to two structures that represent the most distinctive features of this group of funguslike organisms. The two structures (antheridia and oogonia) usually arise as lateral projections from hyphae that extend outward from the substrate upon which the water mold occurs. The oogonium (which can be regarded as the "female" structure) is spherical and delimited from the subtending hypha by a septum. At first, the protoplasm within the oogonium is granular and multinucleate, but eventually most of the nuclei disintegrate and the protoplasm itself condenses around the remaining nuclei to produce oospheres, each with a single nucleus present (PLATE 12). The nuclei originally present in the oogonium are diploid, but meiosis occurs and the nucleus of the oosphere is haploid. In some species of water molds, the oogonium contains only a single oosphere, but in most species the oogonium contains at least several oospheres.

Antheridia are much smaller than oogonia and often occur on the same hypha, sometimes immediately below the oogonium, but in a number of species these structures are produced on different hyphae. The antheridium (which can be regarded as the "male" structure) first takes the form of a thin tube that grows up towards the oogonium. When mature, the terminal portion of the tube (the antheridium proper) becomes delimited by a septum, the same situation that exists for the oogonium. At this point, one or more thin tubes (the fertilization tubes) develop from the antheridium and grow towards and then penetrate the wall of the oogonium to reach the oospheres. Eventually, a single fertilization tube (or a branch there from) makes contact with each oosphere. Nuclei from the antheridium then migrate through the tubes, with one nucleus entering each oosphere and fusing with the nucleus already present.

As was the case for the oogonium, meiosis occurs in the antheridium, so that the nuclei arriving from that structure are haploid. After this fertilization event, which gives rise to a diploid nucleus, a thick wall develops around each oosphere, converting the latter into an oospore. The wall of the oospore is usually smooth, and the entire structure is highly resistant to unfavorable environmental conditions. There is no mechanism for discharging the oospores from the oogonium; they are liberated when the latter disintegrates over time. Ultimately, however, the oospore germinates, potentially to produce a new mycelium. As mentioned earlier, the nuclei found within the mycelium of a water mold are diploid, in contrast to the nuclei of the true fungi which are almost always haploid.

Water Molds as Parasites

As already noted, certain types of water molds are important parasites, and some have had an impact upon human history. The two most prominent examples are undoubtedly *Phytophthora infestans*, the organism responsible for late blight of potatoes, and *Plasmopara viticola*, the cause of the downy mildew of grapes.

Late Blight of Potato

Phytophthora infestans is likely to have been native to South America, where the potato evolved and wild relatives of the plant still exist. In its natural habitats, the organism does not seem to have been a major problem. However, when the fungus somehow made its way to North America in the early 1840s, it caused major outbreaks of the disease that became known as late blight of potato.

The disease first manifests itself when the leaves of an infected plant develop spots along their margins. The spots darken and spread, and are accompanied by the appearance of glistening halos of white mycelia on the lower surface of the leaf. Numerous elongated and branched sporangiophores, with each branch bearing a lemon-shaped sporangium at its tip, arise from the mycelia. The sporangia eventually become detached and can be splashed short distances by raindrops or carried considerable distances by the wind.

Upon coming in contact with another potato plant, the sporangium germinates and gives rise to a new mycelium (and thus point of infection). Late in the year, when temperatures drop, the sporangium gives rise to a group of swimming zoospores. Under moist conditions, when there is a film of water present on the potato plant and the soil is waterlogged, the zoospores can migrate from one plant to another and also can make their way through the soil to infect the potato underground. The disease ultimately affects all parts of the potato plant. The potato itself develops brown streaks that penetrate the flesh and literally turn it into an offensive organic soup that is totally unfit for human consumption.

Although *Phytophthora infestans* caused problems in North America, the impact of the disease in northern Europe was much worse. Unintentionally imported into Ireland about the same time as it was first noticed in North America, the fungus wiped out the Irish potato crop in the damp, cool summers of 1845 to 1847. Because the Irish were largely dependent upon the potato as a source of food, widespread famine resulted. It has been estimated that the disease contributed to more than a million deaths and drove many more to emigrate from Ireland. Many of the displaced individuals came to the United States. Ten years after the first epidemic, the population of Ireland had dropped from eight million down to four million. Unquestionably, this fungus had a significant impact upon human history.

Grapevine Downy Mildew

Plasmopara viticola, the causal agent of grapevine downy mildew, had a somewhat different impact on human history. First of all, the fungus is common on wild species of grape in eastern North America, where it does little harm. When it was accidentally introduced into Europe in the 1870s, it was quite a different story. European grapes had no resistance to *P. viticola* and were quickly devastated. In France alone, millions of people depended upon grapes (and the wine produced from grapes) for a living, so this was certainly no small matter. Fortunately, Alexis Millardet, a professor at the University of Bordeaux, literally stumbled upon the fact that applying a solution of copper sulfate and lime to grape plants was effective in controlling the fungus. Interestingly, the Bordeaux mixture, as it became known, also has the distinction of being one of the world's first practical fungicides.

Plasmopara viticola is similar to *Phytophthora infestans* in that the presence of the organism is first apparent on the leaves of the plant. Young infections appear on the upper surface of the leaf as small, greenish-yellow translucent spots that are difficult to see. Over time, these enlarge and develop into irregular pale-yellow to greenish-yellow spots ⅜ inch (1 cm) or more in diameter. Infected leaf tissue gradually becomes dark brown and brittle. Severely infected leaves eventually turn completely brown, wither, curl, and then drop off.

Plasmopara viticola overwinters in the oospore stage in infected leaves on the ground. The oospores germinate in the spring to produce zoospores, which can be spread by wind and rain. When a zoospore encounters a potential host plant, the former germinates and produces a hypha that enters a leaf through a stoma on the lower surface. Once inside, the fungus grows throughout the tissues of the plant. New areas of infection are usually visible in about seven to twelve days. At night during periods of high humidity and temperatures of at least 55°F (13°C), the mycelium produces microscopic, branched sporangiophores that extend out through the stomata of infected leaves. These make up the delicate, white to grayish white cottony "downy mildew" for which *P. viticola* received its common name. Sporangia are produced on the tips of the sporangiophores, and these can cause secondary infections on other parts of the host plant.

Dieback

It is hard to imagine a water mold eliminating entire forests, but such a thing has actually happened in some parts of the world. The organism responsible, *Phytophthora cinnamomi*, is another member of the same genus as *P. infestans*, which causes the late blight of potatoes. Both species of fungi have been reported from throughout the world, but unlike *P. infestans*, *P. cinnamomi* is not restricted to a single host plant. In fact, *P. cinnamomi* has been recorded from more than 1000 different kinds of plants, including such taxonomically diverse groups as club mosses, ferns, cycads, and conifers in addition to flowering plants. This is a remarkable range for a plant pathogen and underscores the magnitude of the potential threat posed by this particular fungal species.

A plant susceptible to *Phytophthora cinnamomi* becomes infected when a zoospore enters a root just behind the root tip. Zoospores require water to swim through the soil. Consequently, infection is most likely when soils are moist. Once inside the root, the zoospore gives rise to a vegetative mycelium that grows throughout the tissues of the root and lower stem of the host plant, destroying their structure and effectively cutting off the plant's supply of water and nutrients from the soil. This condition gives rise to the common names "root rot" or "dieback" often applied to *P. cinnamomi*. Presence of the fungus often leads to death of the plant, especially in dry summer conditions when plants are likely to be water stressed.

Western Australia is one of the areas of the world where *Phytophthora cinnamomi* has had a devastating effect upon the native vegetation. The disease was first identified as present in Australia in 1930 but was not recognized as the pathogen causing dieback in Western Australia until 1965. By then the disease was firmly established in forests throughout the country. Forests in Western Australia dominated by *Eucalyptus marginata* (commonly called jarrah) were among the most affected by dieback. Jarrah is the single most important timber tree in Western Australia, and the damage caused by *P. cinnamomi* has had a significant impact both economically and ecologically (Cahill et al. 2008).

Damage to forests that was suspected to have been caused by *Phytophthora cinnamomi* was first detected in the United States about two centuries ago, and there is some evidence that it may have killed many American chestnut (*Castanea dentata*) trees well in advance of the chestnut blight fungus (*Cryphonectria parasitica*) that will be discussed in Chapter 11 (Anagnostakis 2001). *Phytophthora cinnamomi* is a major pathogen of avocado in southern California and has devastated forests of ohia lehua (*Metrosideros polymorpha*) in Hawaii. Other trees known to be affected by the disease include shortleaf pine (*Pinus echinata*) and various species of oak (*Quercus*). In Europe, several species of oak, including cork oak, have been affected.

Zoospores are considered the primary source of infections for *Phytophthora cinnamomi*. During periods of warm and rainy weather, the vegetative mycelium produces numerous zoosporangia in which the asexual zoospores develop. When the zoospores are released, they are motile enough to travel short distances through the soil but are unlikely to infect new hosts over anything more than relatively short distances. It is possible for zoospores to be carried across much larger distances by moving water, but this is probably an uncommon event.

During periods when environmental conditions are less favorable, larger, usually thick-walled chlamydospores are produced. Chlamydospores, which are capable of surviving for several years within infected plants and in the soil, almost certainly account for the spread of the disease over long distances through the movement of infected soil and plant material.

Sexual reproduction involving the production of oogonia appears to be exceedingly rare in the field in Australia, but the formation of these structures can be induced in the laboratory, thus confirming that the organism is indeed a water mold.

Class Trichomycetes

In addition to the water molds and chytrids, various representatives from several other groups of fungi also occur in aquatic habitats. Some of these have an important but underappreciated role as decomposers of organic substrates introduced into water. Certain other aquatic fungi are so ecologically and morphologically distinct from other fungi that they have not always been recognized as being fungi.

Members of the phylum Zygomycota are generally regarded as primitive because they do not produce complex fruiting structures and most produce hyphae that lack septa. In the latter respect, they are similar to chytrids and, like the chytrids, were once placed in the assemblage of fungi and funguslike organisms recognized as the class Phycomycetes. However, the spores produced in the Zygomycota are nonzoosporic (that is, they lack flagella).

The phylum Zygomycota consists of two taxonomic classes, the Zygomycetes and the Trichomycetes. As a group, fungi assigned to the Zygomycetes are terrestrial, although many of them tend to be confined to moist places. Nevertheless, because they are essentially terrestrial, these fungi will be discussed in Chapter 3. In contrast, the second class (Trichomycetes) consists of numerous examples that are aquatic.

There is little question that members of the Trichomycetes comprise a rather distinctive group of fungi. In fact, members of the Amoebidiales and Eccrinales, two of the four taxonomic orders once assigned to this class, have been removed because evidence obtained from molecular studies clearly indicates that they are protozoans and not fungi. However, these non-fungal trichomycetes (as the name of the group is used in an informal sense) are still studied by mycologists. As such, this situation is not unlike the one already described for the non-fungal oomycetes. However, only those examples (that is, members of the Asellariales and Harpellales) that belong to the kingdom Fungi are considered in this chapter.

Trichomycetes are found almost exclusively in the digestive tracts (gut) of living arthropods, including insects, crustaceans, and millipedes. The arthropod hosts involved include representatives from freshwater and marine as well as terrestrial habitats. These fungi appear to be cosmopolitan, since they have been found wherever adequate sampling for the group has been carried out. Most species occur in the hindgut of their arthropod host, where they are attached to the lining of the gut by a specialized structure called a holdfast. With the exception of a single species, in which the lining of the gut is penetrated to reach the underlying tissues, these fungi do not appear to disrupt their hosts. This seems to be case even in those instances in which a particular host may have so many of the fungi present that the inside of the gut has a fuzzy or hairy appearance. Indeed, the name trichomycetes, which literally translates as "hair fungi," is derived from the appearance of these fungi in mass.

Depending upon the species involved, the minute thallus of a trichomycete may be branched or unbranched. In some examples, the individual hyphae are regularly septate, while in others hyphae lack septa except at the bases of reproductive cells. Other differences that exist among members of this class are the size and shape of the spores, the number of appendages on the spores, the number of spores borne per branchlet, the shape and nature of the holdfast, and the type of arthropod host (PLATE 13). The latter is a factor because particular species often exhibit varying degrees of host specificity, often being restricted to the members of certain genera or families of arthropods.

Sexual reproduction in the trichomycetes is not well documented, although there is evidence that it does occur in some species. As such, reproduction in the group apparently occurs largely by asexual spores. In the members of one order (Harpellales), special structures called trichospores are produced. Each trichospore actually consists of a sporangium containing a single spore and having one to several basally attached filamentous appendages. The entire structure functions as a unit for dispersal, and the appendages apparently cause the dispersed spore to adhere more readily to bits of organic debris upon which a potential host might feed. If ingested by a suitable host, the trichospore germinates quickly, and the developing thallus attaches to the lining of the gut.

Since trichomycetes do not seem to be parasites, the question might be asked as just what type of association of host and fungus is involved. Many trichomycetes appear to be commensalistic (that is, the fungus benefits but the host is not affected), but under certain circumstances, fungi may provide their hosts with essential organic nutrients that would otherwise be unavailable. Moreover, at least one species (*Smittium morbosum*) is lethal to mosquito larvae, and several species found in blackflies (Simuliidae) and other insects are known to invade the ovary of the host and produce cysts that are "oviposited" by females in lieu of eggs. The cysts would reduce overall fertility in populations of the aquatic insects involved.

Aquatic Members of Phylum Ascomycota

Mitosporic ascomycetes are members of the phylum Ascomycota in which only asexual spores are produced. Most of these fungi are terrestrial and will be discussed in Chapter 3. However, two ecological assemblages of mitosporic ascomycetes—the aquatic hyphomycetes and the aero-aquatic fungi—are among the more distinctive and underappreciated fungi found in aquatic habitats.

Aquatic Hyphomycetes

Aquatic hyphomycetes were first recognized as a distinct ecological assemblage by the British mycologist Cecil Terence Ingold (1942), and therefore they are often referred to as Ingoldian hyphomycetes. These fungi are saprotrophs on dead leaves, twigs, and other types of plant material in water. They are particularly abundant on submerged, decaying leaves of broadleaf trees, especially when the leaves occur along the bottom of a well-aerated mountain stream. The mycelium of the fungus proliferates throughout a leaf, ultimately breaking it down.

Studies have shown that aquatic hyphomycetes are the organisms primarily responsible for degrading leaves in streams. As such, these fungi are exceedingly important to the overall ecology of the streams, since dead leaves represent the single major source of plant organic material for most streams. One way in which aquatic hyphomycetes are important to other stream inhabitants will be described below.

As the mycelium invades a leaf, it eventually produces numerous specialized spore-producing hyphae that extend out into the surrounding water. The spores themselves are truly distinctive. Each spore is branched, and the most common expression is to have four branches (or appendages) diverging from a single point, a condition that is referred to as tetraradiate (PLATE 14).

In a well-aerated stream supplied with a layer of dead submerged leaves, enormous numbers of such spores are liberated, often several thousand per quart (0.95 liter) of water. The spores tend be trapped in the small pockets of foam or froth that often develop at the bases of ripples in the stream, and if a sample of the foam is collected, placed on a slide, and examined under a microscope, examples of tetraradiate spores are easily observed.

The tetraradiate spore shape is thought to be an adaptation that allows a more stable attachment of the spore to potentially suitable substrates that are encountered in flowing water. Any spore with a more or less spherical shape (which applies to the majority of spores) would have little chance of coming to rest on a submerged leaf subjected to water currents, while the appendages of a tetraradiate spore would function like a miniature anchor and thus enable the spore to remain attached to the surface of the leaf once contact has been made.

Stream Ecology and Food Chains

Dead leaves falling into streams have long been recognized as an important food source for many aquatic insects and other invertebrates. As they feed, some of these organisms (referred to as shredders) fragment the leaf material into small particles that can be utilized by other stream organisms. Shredders constitute an ecological and not a taxonomic assemblage, although members of three orders of insects (caddisflies, stoneflies, and true flies) are the dominant groups involved.

As noted above, dead leaves submerged in a stream are quickly invaded by aquatic hyphomycetes, and those leaves that have been partially degraded ("conditioned") by these fungi are the ones upon which the shredders are most likely to feed. In fact, studies have demonstrated that shredders are selective in their foraging behavior. In other words, they prefer some aquatic hyphomycetes over others.

Because the mycelia of the fungi are present throughout the leaf, the actual material upon which shredders feed consists of leaf tissue and fungal hyphae. This combination is nutritionally richer for the insect than the leaf tissue alone. First, some of the indigestible (to the insect) substances in the leaf have been converted into digestible molecules by the enzymes produced by the fungus. Second, the hyphae of the fungus have a higher food quality than the tissues of the leaf. Although the average trout fisherman is unlikely to be aware of this fact, aquatic hyphomycetes actually play an essential role in the food chains of small streams. These food chains ultimately include the trout.

Aero-aquatic Fungi

The aero-aquatic fungi also occur on dead leaves, twigs, and other types of dead plant material submerged in water, but the body of water involved is more likely to be a stagnant pool than the well-aerated stream described for aquatic hyphomycetes. Another difference is that so long as the mycelium of an aero-aquatic fungus is below the surface of the water, it does not form asexual reproductive structures. As noted above, this is in marked contrast to the aquatic hyphomycetes, in which the mycelium does produce asexual reproductive structures (spores) while completely submerged. If the water level drops and the mycelium of an aero-aquatic fungus is exposed to air under moist conditions, it then does form asexual reproductive structures. The asexual propagules produced are too complex to be considered spores but have the same function.

These asexual propagules of aero-aquatic fungi are relatively large and in their own way just as distinctive as the spores of aquatic hyphomycetes. Each propagule consists of either a spherical network enclosing an open space or a tightly coiled and hollow helical structure (PLATE 15). Both types trap air inside and can float on the surface of the water once liberated. The unique configuration of these structures appears to be an adaptation for dispersal by water.

Aquatic Macrofungi

Relatively few macrofungi are aquatic, but the fruiting bodies of *Mitrula paludosa* (swamp beacon) often arise above the surface of the water from submerged well-decayed wood and mats of dead leaves found at the edges of shallow pools, in *Sphagnum* bogs, or even roadside ditches. *Mitrula paludosa* (PLATE 16) belongs to the phylum Ascomycota, a group that will be discussed in some detail in Chapters 3 to 5. The fruiting body, which is no more than ¾ to 2 inches (2 to 5 cm) tall, consists of a slender light-colored stalk with an expanded bright orange-yellow upper portion. Because of its bright color, the swamp beacon is an easy fungus to spot in nature.

3
The Most Ubiquitous of All Fungi

Phylum Zygomycota 47

Sexual and Asexual Reproduction in Zygomycetes 47

Zygomycetes on Dung 49

Zygomycetes as Parasites of Insects 51

Phylum Ascomycota 53

Asexual Reproduction in Ascomycetes 53

Anamorph and Teleomorph Stages 54

Septate Hyphae 55

Asexual Spores 55

Classification of Ascomycota 56

Subphylum Pezizomycotina 57

Aspergillus and *Penicillium* 58

Identifying Terrestrial Mitosporic Ascomycetes 60

Yeasts 61

Ascomycetes—The Best-known Yeasts 62

Basidiomycete Yeasts 63

Other Fungi That Produce Yeastlike Cells 64

Yeasts in Nature 64

The vast majority of fungi produce spores on or within fruiting structures that are too small to be observed in any kind of detail without the use of a microscope. Moreover, the vegetative body of such fungi tends to be either limited in extent or not easily detected as a result of being highly dispersed or more or less completely immersed within a particular substrate. These terrestrial microfungi, an assemblage that encompasses representatives from all major taxonomic groups, are truly ubiquitous in nature, where they are of considerable (albeit often little appreciated) ecological importance in all types of terrestrial habitats. All too often, when terrestrial microfungi are noticed, it is because of their negative aspects (PLATE 17).

Members of two phyla (the Zygomycota and the Ascomycota) undoubtedly make up the majority of terrestrial microfungi. As indicated in Chapter 2, the Zygomycota is made up of two classes, the Zygomycetes and the Trichomycetes, the second of which was discussed in the context of aquatic microfungi.

Phylum Zygomycota

Members of the Zygomycota (or zygomycetes if considered in the less formal sense) are usually rather inconspicuous, but they are often exceedingly common in soil and on the dung of animals as well as also occurring on many other types of substrates, some of which are very familiar to most people. For example, if one encounters moldy bread or a piece of fruit that has gone bad, it is very likely that a member of this group of fungi is involved. In fact, *Rhizopus stolonifer*, one of the most ubiquitous of all zygomycetes, is often referred to as bread mold (PLATE 18).

Most zygomycetes are saprotrophs, but some species are parasites or pathogens of plants, animals (including humans), and even other fungi. The latter is the case for *Spinellus fusiger* (bonnet mold), which parasitizes the fruiting bodies produced by certain members of the Basidiomycota (PLATE 19).

Sexual and Asexual Reproduction in Zygomycetes

The most important feature defining the zygomycetes is the production of a thick-walled sexual spore called a zygospore. The latter develops within a special structure (the zygosporangium) that is formed by the fusion of two equal or unequal gametangia (singular: gametangium). Each gametangium arises from a vegetative mycelium, either the same mycelium or two different mycelia. In some of the more typical zygomycetes, the mycelium is a well-developed system of coenocytic hyphae for which some degree of differentiation is apparent in the production of groups of rootlike rhizoids, rapidly growing aerial hyphae (stolons) that are responsible for extending the mycelium to new areas of the substrate upon which it occurs, and sporangiophores that project above the substrate and develop a terminal sporangium containing numerous asexual spores.

When the wall of the sporangium is disrupted, these spores are liberated into the air. Because of their small size, the spores can easily remain aloft long enough to reach suitable new substrates. In fact, it is usually safe to assume that the spores of *Rhizopus stolonifer* (mentioned above) are constantly present in most indoor environments. A moistened slice of bread exposed to the air for a few minutes and then placed in a plastic bag for a few days will almost invariably develop fuzzy spots that are likely to be the developing mycelia of this fungus.

In some zygomycetes, most notably members of the genus *Mucor*, the spores contained in the sporangia are not dispersed in the air. Instead, the wall of the sporangium dissolves away except for a basal cup-like collar, allowing moisture to seep into the mass of spores and converting them into a sporangial drop. This mass is dispersed when it comes into contact with some passing arthropod or is displaced by a falling raindrop.

In zygomycetes that produce asexual spores in the manner described above, the organism can persist for extended periods of time by simply repeating the cycle of going from spore to mycelium to spore again. Occasionally sexual reproduction does occur. As already noted, sexual reproduction in these fungi involves the production of gametangia. Although it would be tempting to designate the two different gametangia as male and female, they usually cannot be distinguished morphologically. Instead, mycologists simply refer to them as "+" and "−" strains (or mating types).

The first step in the formation and ultimate fusion of two gametangia is the production of special hyphae called zygophores. Compatible zygophores are attracted to one another and fuse at the point of contact. The tips of the two zygophores are then delimited by septa from their subtending hypha. At this point, each tip is a separate cell, called a progametangium at first but then developing into a functional gametangium. The two gametangia fuse and their contents are mixed together in a single cell, known as the prozygosporangium. This structure develops a thick wall around itself and becomes a zygosporangium.

In older texts, the zygosporangium is often referred to as a zygospore, but studies have clearly demonstrated that the two structures are not the same. In fact, the zygospore, which ultimately gives rise to a new mycelium, is produced by the zygosporangium. This does not take place immediately, and the zygosporangium can be considered as a resting stage in the life cycle of the fungus.

Plate 1. Mycelium of a fungus in a decaying log. Photograph by John Plischke.

Plate 2. The bizarre fruiting body of *Aseroë rubra* (starfish stinkhorn). This specimen was found in Hawaii, where the fungus seems to have been introduced. Photograph by Taylor Lockwood.

Plate 3. Teliospores of *Phragmidium mucronatum* (rose rust). The microscopic structures produced by fungi are often quite intricate and sometimes beautiful. Photograph by Jerry Cooper.

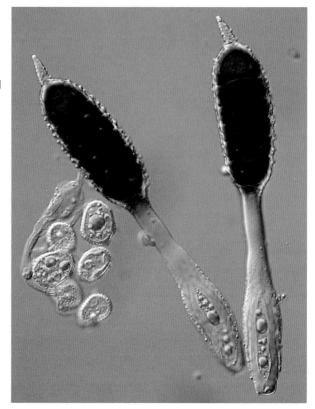

Plate 4. *Agaricus campestris* (meadow mushroom). This species, which occurs in fields and other grassy areas, is very similar to *A. bisporus* (cultivated mushroom). Photograph by Alan Bessette.

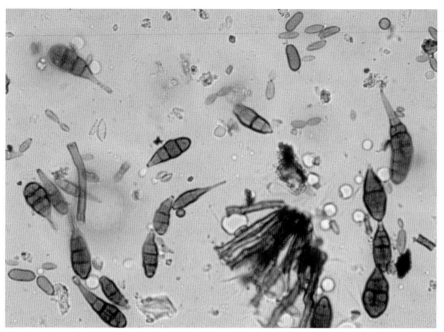

Plate 5. Airborne fungal spores obtained from outdoor air. The sample was collected in October, when few pollen grains are in the air. The club-shaped spores are those of *Alternaria*, an anamorphic (see Chapter 3) member of the phylum Ascomycota. Photograph by Estelle Levetin.

Plate 6. Chytrids on the filament of a green alga. Photograph by Peter Letcher.

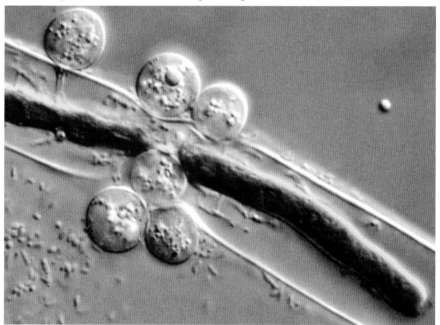

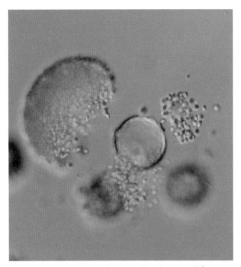

Plate 7. Zoospores being discharged from the zoosporangium of a chytrid. Photograph by Peter Letcher.

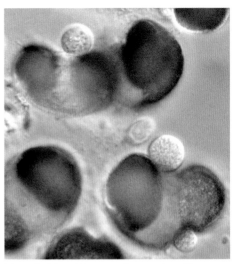

Plate 8. Chytrids on pollen grains. Pine (*Pinus*) pollen grains are often used to "bait" for chytrids, which often appear on such substrates within a few days. Photograph by Peter Letcher.

Plate 9. Spherical bodies in skin cells from a northern leopard frog (*Rana pipiens*). The presence of these structures indicates that the frog is infected by *Batrachochytrium dendrobatidis*. Photograph by Joyce Longcore.

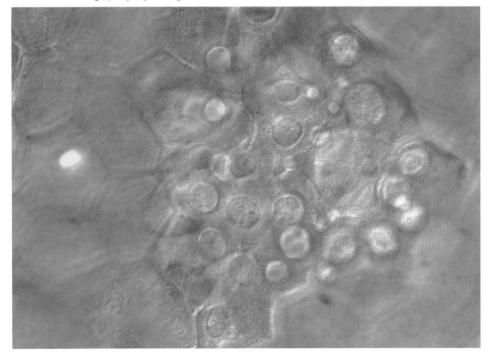

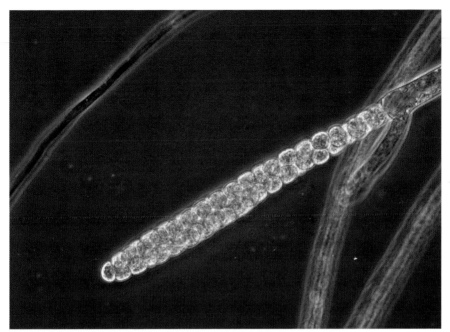

Plate 10. Cigar-shaped zoosporangium of a water mold. The zoosporangium contains fully mature zoospores ready to be discharged. Photograph by David Padgett.

Plate 11. Zoosporangium of *Saprolegnia* in which two secondary zoosporangia have developed. The first of the two secondary zoosporangia is empty, and the second is still forming. Photograph by David Padgett.

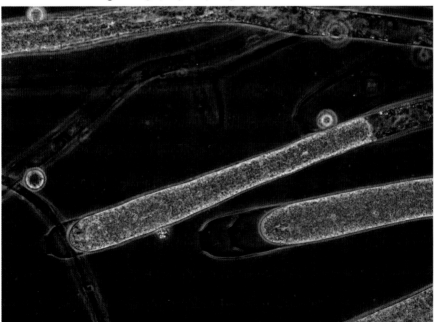

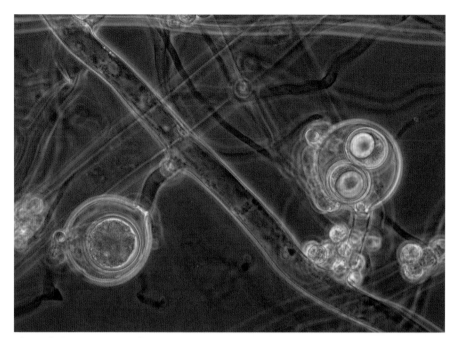

Plate 12. Two oogonia of a water mold. The example on the left has one oosphere present, while the example on the right contains two mature zoospores. Photograph by David Padgett.

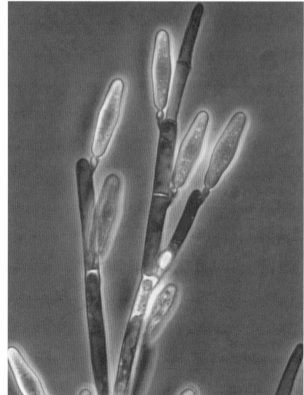

Plate 13. The minute thallus of a trichomycete (*Smittium simulii*) collected from the digestive tract of a midge (Chironomidae) larva. Photograph by Merlin White.

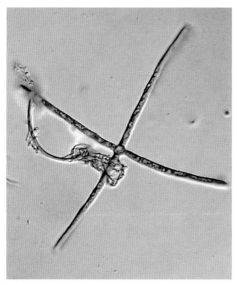 

Plate 14. Tetraradiate spore of an aquatic hyphomycete (*Lemonniera centrosphaera*). Photograph by Jerry Cooper.

Plate 15. Asexual propagule of an aero-aquatic fungus (a member of the genus *Helicoön*). Photograph by Jerry Cooper.

Plate 16. Fruiting bodies of *Mitrula paludosa* (swamp beacon) commonly occur on dead leaves in wet areas. Photograph by Martin Schnittler.

Plate 17. Mycelium of a terrestrial microfungus (*Penicillium digitatum*) on an orange. Photograph by Emily Johnson.

Plate 18. Sporangiophores of *Rhizopus stolonifer*. This species is one of most common and widespread all fungi. Photograph by Alena Kubátová.

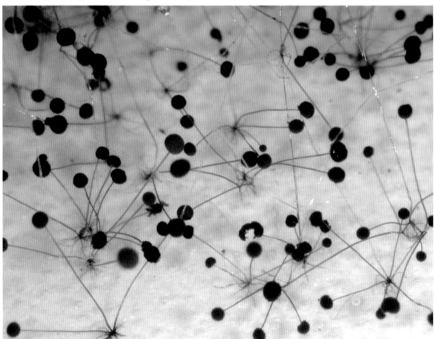

Plate 19. *Spinellus fusiger* (bonnet mold). The sporangiophores of this fungus radiate outward from the cap of a member of the genus *Mycena* in the Basidiomycota.
Photograph by Emily Johnson.

Plate 20. Sporangiophores of *Pilobolus* (hat-thrower). This fungus commonly occurs on the dung of large herbivores.
Photograph by George Barron.

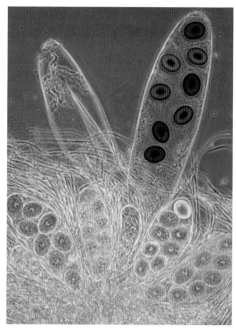

Plate 21. Dead fly infected by *Entomophthora muscae*. Photograph by Daniel Mahr.

Plate 22. Ascus and ascospores of *Ascobolus calesco*. This species is an example of a fungus that occurs on dung. Photograph by Dan Mahoney.

Plate 23. Conidiophore of *Aspergillus*. The major distinguishing feature of this genus is a terminal vesicle with radiating phialides. Interestingly, the name for this genus is based on the resemblance of the conidiophore to a holy water sprinkler (or aspergillum). Photograph by Keith Seifert.

Plate 24. Conidiophore of *Penicillium*. Species of *Penicillium* are usually easy to recognize as a result of their dense, brush-like conidiophores. Photograph by Keith Seifert.

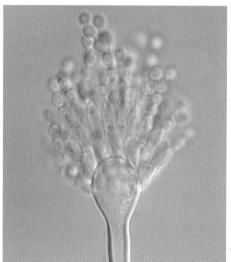

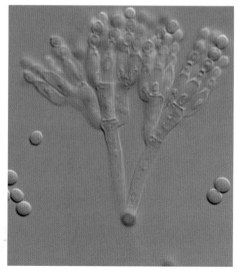

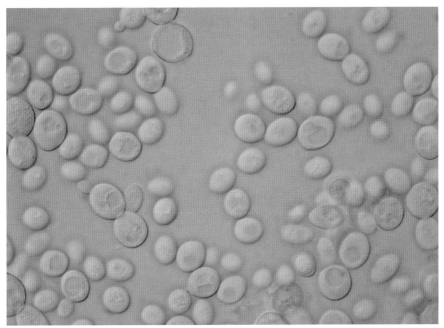

Plate 25. *Saccharomyces cerevisiae* (baker's yeast or brewer's yeast). Several of the cells are in the process of budding.
Photograph by Lori Carris.

Plate 26. The fruiting bodies of *Cordyceps robertsii* (vegetable caterpillar fungus) arise from the bodies of caterpillars that have been infected by the fungus.
Photograph by Wang Yun.

Plate 27. Fruiting bodies of *Cyttaria gunnii* (beech strawberry) on the branch of a silver beech (*Nothofagus menziesii*) in New Zealand. Photograph by Peter Johnston.

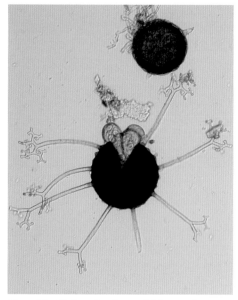

Plate 28. Cleistothecium of *Microsphaera penicillata*, a powdery mildew found on the leaves of lilac (*Syringa vulgaris*) during the late summer and early fall. Photograph by Lori Carris.

Plate 29. *Chlorociboria aeruginascens* (green stain of wood). The mycelium of this fungus stains the wood in which it occurs a bluish green. Photograph by Emily Johnson.

Plate 30. Fruiting bodies of *Leotia lubrica* (jelly club) are found on soil or forest floor litter. Photograph by Emily Johnson.

Plate 31. Fruiting bodies infected by the parasitic fungus *Hypomyces lactifluorum*. Because of their color, such fruiting bodies are often referred to as "lobster mushrooms." Photograph by Emily Johnson.

Plate 32. The thallus of a member of the order Laboulbeniales is unlike anything else found in the kingdom Fungi. Photograph by Alex Weir.

Plate 34. Perithecium of *Podospora fimiseda* on dung. Scattered bristlelike hyphae cover the outer surface of the perithecium. Photograph by Dan Mahoney.

Plate 33. *Apiosporina morbosa* (black knot of cherry). The conspicuous black, knotlike structures on the branches of a cherry tree are produced when latter has been infected by this fungus. Photograph by Emily Johnson.

Plate 35. *Morchella esculenta* (common yellow morel). The honeycomb-like cap of the common yellow morel is distinctive and makes this fungus easy to recognize. Photograph by Emily Johnson.

Plate 36. Two fruiting bodies of *Morchella elata* (black morel). This species is often the first morel to appear in the spring. Photograph by Emily Johnson.

Plate 37. *Gyromitra brunnea* (brown false morel). Unlike the true morels, false morels can be deadly poisonous and should not be collected for human consumption. Photograph by Emily Johnson.

Plate 38. *Tuber melanosporum* (Périgord black truffle) is generally regarded as the most exquisite of all truffles. Photograph by Ian Hall.

Plate 39. Species of *Elaphomyces*, including *E. granulatus* (deer truffle), are the most common hypogeous fungi in temperate regions of the world. Photograph by John Plischke.

Plate 40. *Sarcoscypha occidentalis* (stalked scarlet cup). The bright scarlet red fruiting bodies are relatively conspicuous in spite of their small size. Photograph by Emily Johnson.

Plate 41. *Scutellinia scutellata* (eyelash cup). The radiating long dark hairs around the rim of the cuplike fruiting body make this an easy fungus to identify. Photograph by Ron Wolf.

Plate 42. Apothecium of *Ascobolus* on dung. The tips of the projecting asci are oriented in the direction of incoming light. Photograph by Dan Mahoney.

Plate 43. *Xylaria polymorpha* (dead man's fingers). The fingerlike fruiting bodies of this fungus make it one of the more distinctive members of the family Xylariaceae. Photograph by Emily Johnson.

Plate 44. Fruiting bodies of *Hypoxylon fragiforme* are restricted largely to dead branches of beech (*Fagus*). Photograph by Emily Johnson.

Plate 45. *Daldinia concentrica* (carbon balls) on a decaying log. Note the concentric zones of alternating light and dark bands in the interior of the fruiting body that has been broken open. Photograph by Emily Johnson.

Plate 46. *Diatrype virescens*. The yellow-green color of the wart-like stromata of this species is distinctive. Photograph by Randy Darrah.

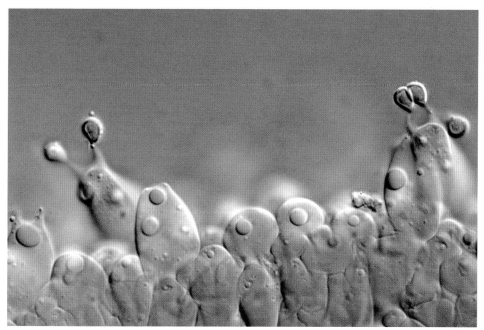

Plate 47. Basidia and developing basidiospores in the hymenium of a member of the Basidiomycota. Photograph by Yuri Novozhilov.

Plate 48. *Amanita bisporigera* (destroying angel), one of the most deadly poisonous of all fungi. Photograph by Emily Johnson.

Plate 49. Species of *Russula* often have brightly colored caps and a stalk that is brittle and breaks in much the same way as a piece of chalk. Photograph by Emily Johnson.

Plate 50. Brightly colored fruiting bodies of *Hygrocybe cuspidata* (scarlet waxy cap). Photograph by Emily Johnson.

Plate 51. *Marasmius siccus* (orange pinwheel marasmius). This fungus commonly occurs on dead leaves. Photograph by Emily Johnson.

Plate 52. Fruiting bodies of *Mycena haematopus* (bleeding mycena) produce a watery exudate when the stalk is broken. Photograph by John Plischke.

Plate 53. *Pluteus cervinus* (deer mushroom) is a common and widespread example of an agaric with pink spores. Photograph by Emily Johnson.

Plate 54. *Coprinus comatus* (shaggy mane). This fungus is edible if the fruiting bodies are collected while still fresh. Photograph by Emily Johnson.

Plate 55. *Boletus edulis* (king bolete) is one of the most highly prized edible fungi. Photograph by Alan Bessette.

Plate 56. *Leccinum insigne* (aspen scaber stalk), like other species in the genus *Leccinium*, are immediately recognizable because of the presence of conspicuous tufts of dark hyphae (scabers) on the stalk. Photograph by Emily Johnson.

Plate 57. *Laetiporus sulphureus* (sulfur shelf) is one of the more colorful and better known of the polypores. It often forms large fruitings that can be spotted from a considerable distance away. Photograph by Bill Roody.

Plate 58. *Trametes versicolor* (turkey tail). This exceeding common fungus is one of the more colorful polypores. Photograph by Denise Binion.

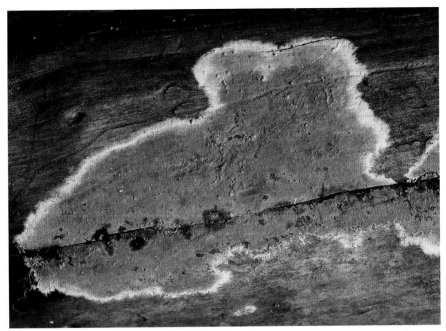

Plate 59. *Coniophora puteana* (butt rot). The resupinate fruiting body of this fungus extends over the surface of the substrate upon which it occurs. Photograph by Emily Johnson.

Plate 60. Chanterelles, including *Cantharellus lateritius* (smooth chanterelle), are highly prized as edible fungi. Photograph by Emily Johnson.

Plate 61. *Hericium erinaceus* (bearded tooth). The fruiting body of this fungus, which has the general appearance of frozen waterfall, is unlikely to be confused with that of any other fungus. Photograph by Ron Wolf.

Plate 62. Example of a particularly colorful coral fungus. Photograph by Emily Johnson.

Plate 63. *Lycoperdon perlatum* (gem-studded puffball). This is one of the more common and widely distributed puffballs. Photograph by Emily Johnson.

Plate 64. *Scleroderma citrinum* (common earthball). The fruiting bodies of the common earthball resemble those of puffballs but differ in having a rough, much thicker peridium. Photograph by Emily Johnson.

Plate 65. *Geastrum saccatum* (round earthstar). Widely distributed throughout North America, the earthstars are some of the most distinctive of all fungi. Photograph by Emily Johnson.

Plate 66. *Cyathus striatus* (bird's nest fungus). This species often fruits on mulch beds after a period of rainy weather. Photograph by Emily Johnson.

Plate 67. *Dictyophora multicolor* (bridal veil fungus). The netlike skirt that is the defining characteristic of the genus *Dictyophora* is usually well developed in this species. Photograph by Taylor Lockwood.

Plate 68. *Tremella mesenterica* (witches' butter). The fruiting bodies of this jelly fungus take the form of convoluted gelatinous masses. Photograph by Bill Roody.

Plate 69. *Pseudohydnum gelatinosum* (jelly tooth). This fungus has a combination of features (jellylike but with the basidiospores produced on toothlike spines) not known for any other fungus. Photograph by Emily Johnson.

Zygomycetes on Dung

Some of the more distinctive zygomycetes belong to the family Pilobolaceae. These fungi, in spite of their relatively simple structure, discharge their spores in an amazing way. Species of *Pilobolus* (hat-thrower) develop with considerable regularity upon the freshly deposited dung of large herbivorous animals such as sheep, cattle, and horses. Often, after three or four days, the surface of the dung becomes covered with a miniature forest consisting of numerous projecting sporangiophores that develop from vegetative mycelia occurring throughout the dung.

Each sporangiophore arises from a basal bulblike trophocyst that has been cut off from the mycelium proper by the formation of septa. The mature sporangiophore consists of the trophocyst, a slender unbranched main axis $1/4$ to $3/8$ inch (0.5 to 1 cm) in length, and a swollen, subsporangial vesicle located directly below a somewhat flattened, black sporangium containing numerous spores. Because of the septa that delimit the trophocyst from the rest of the mycelium, the entire sporangiophore is essentially a single large cell.

After reaching maturity, osmotically active compounds within the sporangiophore cause the internal pressure to build up to such an extent that it can exceed more than 100 pounds per square inch (7 kg per cm^2). It is common to see a series of small droplets along the length of the sporangiophore that form because of its extremely turgid condition (PLATE 20). The latter eventually causes the subsporangial vesicle to rupture along a line of weakness just below the sporangium. When this occurs, the sporangium is forcibly discharged directly upwards to a distance of as much as 6 feet (2 m). This is an amazing achievement for a sporangiophore that is no more than $3/8$ inch (1 cm) tall. A rough equivalent would be a human throwing a football more than 300 yards (274 m). Some of the rather gluelike mucilaginous contents of the subsporangial vesicle are discharged along with the sporangium, and these allow the latter to adhere to whatever surface is encountered at the end of its flight. However, there is more to this than might be expected.

The sporangium of *Pilobolus* is covered in crystals of calcium oxalate. In addition to serving as a protective layer, the crystals are highly hydrophobic (water avoiding) and cause the sporangium to flip over onto its sticky lower surface after landing in a drop of dew or a film of water that is likely to be present on a leaf or other plant surface. This greatly increases the chances of the sporangium remaining securely in place until consumed (along with the plant it is on) by some other herbivore. The spores of *Pilobolus* pass through the digestive tract of the latter unharmed. In fact, the conditions to which the spores are exposed while in the digestive tract actually stimulate their subsequent germination in the dung when this is deposited.

Although the fact that the sporangium of *Pilobolus* is forcibly discharged is remarkable in itself, this is not the complete story. The fungus has the capability of determining the exact direction in which the sporangium is discharged. How does this take place? First of all, the structure of the upper portion of the sporangiophore is rather like that of a simple "eye" consisting of a clear, light-directing "lens" (the subsporangial vesicle itself) and an orange "retina" (a thickened ring of pigment-containing protoplasm at the base of the subsporangial vesicle). The retina, which is clearly visible with the unaided eye, controls the growth of the sporangiophore very precisely. If the retina is evenly illuminated by light focused by the lens, the growth of the sporangiophore is simply directed upward. If the retina is not evenly illuminated, this elicits a rapid growth response in the main axis of the sporangiophore. This growth response, which occurs just below the subsporangial vesicle, results in the upper portion of the sporangiophore being redirected until the retina is evenly illuminated.

As such, the sporangiophore is strongly positively phototrophic, and when the sporangium is discharged, it is propelled in the direction of the light. This has a considerable advantage for the fungus. By directing the sporangium towards the open area from which the light is coming, there is an excellent chance that it will land within the vegetation beyond the dung. There is certainly no advantage for the sporangium to land on the dung itself, since this is a substrate that has already been exploited by *Pilobolus*. Moreover, the fact that the sporangium can be propelled a considerable distance from the dung represents yet another advantage. Herbivores do not feed upon their own dung, and this means that they are much less likely to consume plants located immediately adjacent to a deposit of dung than those plants located some distance from it.

The remarkable accuracy of this system can be easily demonstrated by placing some fresh dung with *Pilobolus* present in an open culture dish and covering the dish with a pasteboard box or some other similar opaque container. If a pinhole is made in the top of the box, this creates a tiny opening that is the only source of light for the fungus. In actively growing *Pilobolus*, a fresh crop of sporangiophores appears each morning, and these usually discharge their sporangia about the middle of the day. If the box is removed after having remained in place through one of these cycles and the area around the pinhole on the inside of the box is examined, a tight cluster of discharged sporangia will be centered on this tiny opening.

Interestingly, the forcible discharge mechanism of *Pilobolus* is exploited by certain parasitic nematodes, including lungworms in the genus *Dictyocaulus*. These organisms literally "hitch" a ride on the sporangia. The way this works is that larval lungworm nematodes excreted by infected animals such as deer, elk, cattle, and horses climb up a sporangiophore of *Pilobolus* and are discharged along with the sporangium. The larval lungworm nematodes complete their life cycle when they and the sporangium of their *Pilobolus* vector are consumed by a new host. Using a fungus for transportation certainly is not commonplace in nature, but this is one instance in which it does occur.

All of the zygomycetes mentioned thus far produce reproductive structures that are unbranched and appear to be relatively simple structures. Obviously, in the case of *Pilobolus*, what might appear to be simple is actually quite sophisticated, but this fact is not readily apparent. Nevertheless, there are other zygomycetes for which the reproductive structures are intricately branched. Some of these fungi have a rather distinctive appearance.

Zygomycetes as Parasites of Insects

Some of the other more interesting zygomycetes are found in the order Entomophthorales. These fungi are best known as parasites of insects, but some species are parasitic on other kinds of organisms, including certain types of plants. The most familiar example of a fungus that belongs to this group is *Entomophthora muscae*. *Entomophthora* means "insect destroyer" and, as will be described below, the name is certainly appropriate. This species and related forms are usually referred to as entomogenous fungi.

In late summer and autumn, it is not unusual to notice a dead fly hanging from a flower stalk, a blade of grass, or a window pane in a house. Such flies usually hang from their mouth parts and have a puffy, white-striped abdomen or a white halo encircling the body. If these things are observed and the fly in question is the common house fly

(*Musca domestica*) or some related species, it was most likely killed by *Entomophthora muscae*. This fungus reaches a fly as an airborne asexual spore, and only adult flies are subject to infection.

When a spore, which is covered by a mucilaginous substance, comes into contact with the body of a fly, it adheres to the outer surface of the exoskeleton. Soon thereafter, the spore germinates and produces a hypha that penetrates the cuticle of the fly to reach the interior of the body. Penetration often occurs through one of the many intersegmental membranes between the harder parts of the exoskeleton and not through the exoskeleton itself. Fungal hyphae grow throughout the body of the fly, becoming extensive enough to distend the abdomen by the time the fly dies. As the body is distended, the light-colored membranes between the darker and harder portions of the exoskeleton are exposed, giving the fly the characteristic striped appearance.

Once inside the fly, hyphae grow into the fly's brain, causing a distinct change in its behavior, often referred to as summit disease. Instead of acting normally, the fly crawls upwards as high as possible, going to the tip of the branch, flower, stem, or leaf that it is on. This abnormal behavior is accompanied by the formation of specialized fungal structures and gluelike substances that are secreted by the hyphae for attachment. In many instances, the infected fly securely attaches itself by its extended mouthparts to whatever surface it happens to be on, where it may remain for several weeks. Before the fly dies, it spreads out its legs, stretches opens its wings above the thorax, and angles the abdomen away from the surface (PLATE 21). The elevated position that the fly reaches before dying and the distinct posture that the body assumes just before death greatly improve the chances of the spores produced by the fungus actually infecting new hosts.

Shortly after death, large numbers of sporangiophores erupt from the body of the fly, giving the fly a fluffy white appearance. When the sporangiophores are mature, the spores at their tips are forcibly discharged to a distance of as much as $3/4$ inch (2 cm). These discharged spores often form a halo around the body of the fly. Interestingly, most infected flies die a few hours before midnight, when atmospheric humidity is typically high. This timing improves the chances that the fungus will have the right environmental conditions to produce spores.

In late autumn, some infected flies do not produce sporangiophores and spores when they die. Instead, these flies drop to the ground, and the mycelium in their body produces thick-walled resting spores. These survive the winter in the soil and during the following spring produce the spores that can begin a new infection cycle.

Particular species of entomogenous fungi tend to infect specific insects. As already noted, *Entomophthora muscae* is restricted to the common house fly and its close relatives. The more distantly related *E. syrphi* infects only syrphid flies (hover flies or flower flies) and does not affect houseflies. *Entomophthora maimaiga* infects the gypsy moth (*Lymantria dispar*), while certain other species affect only aphids. Because of this specificity, some efforts have been made to utilize entomogenous fungi for biological control of particular insects, with the gypsy moth representing one of the best-known examples. There has been some apparent success, but consistent results have yet to be achieved and this area of research warrants additional study.

Phylum Ascomycota

With more than 30,000 described species, the Ascomycota is the largest phylum of fungi, and by any type of reckoning, they are an extremely significant and successful group of organisms. The morphological variation found within the Ascomycota (or the ascomycetes in a less formal sense) is amazingly diverse, ranging from unicellular forms to those that have complex fruiting bodies. The single feature that holds the group together is a special type of reproductive hypha called an ascus (from a Greek word meaning "sac" or "wineskin").

There are generally eight sexual spores (called ascospores) per ascus (plural: asci), and in the more commonly encountered situation they are lined up like peas in a pod (PLATE 22). Some of the fungi assigned to the ascomycetes do not reproduce sexually and thus do not form asci or ascospores. These fungi have been placed in the ascomycetes on the basis of morphological and/or physiological similarities to ascus-producing taxa as well as by phylogenetic comparisons of DNA sequences. The latter has become exceedingly important over the past few decades.

Asexual Reproduction in Ascomycetes

Ascomycetes that never reproduce sexually or at least are not known to produce asci have been referred to as deuteromycetes or imperfect fungi. Both terms have been used in a formal taxonomic sense, since the fungi involved have been circumscribed as a separate artificial phylum, the Deuteromycota or Fungi Imperfecti, in most earlier systems of classification. (The latter name was based on the idea that these fungi were "imperfect" because they did not reproduce sexually as did the "perfect" sexually reproducing fungi that could be placed in the other phyla.) Today, most mycologists have discarded this concept completely, and data from molecular analyses have been used to place the fungi formally relegated to the Deuteromycota among ascus-bearing groups (if the molecular data reveal that they are ascomycetes) or among other phyla such as the Basidiomycota (if a relationship to this phylum is demonstrated).

Since thousands if not even more taxa have yet to be subjected to molecular studies, there are many fungi for which the exact placement is not yet known. These fungi are sometimes referred to as the mitosporic ascomycetes because they produce only asexual spores and not ascospores in asci. The use of this term has no taxonomic connotation except to assume that most of the taxa involved are ascomycetes, which is generally thought to be the case.

Anamorph and Teleomorph Stages

The literature of mycology often refers to "anamorph," "teleomorph," and "holomorph." These terms, first proposed by Grégoire Hennebert of the University of Louvain and Luella Weresub of Agriculture Canada (1977) and now widely accepted by mycologists, designate the asexual (anamorph) stage of a particular fungi, the sexual (teleomorph) stage of the same fungus, and the whole fungus (holomorph) when considered as a single taxonomic entity.

Although most fungi appear to be capable of reproducing both sexually and asexually, the structures involved in each type of reproduction are not typically produced at the same time. When this is the case, there is often no easy way of knowing whether or not two fungi are the same or different taxa. In many instances, each of the two stages (anamorph and teleomorph) has its own taxonomic name, which was given to it by mycologists who did not realize that they were actually dealing with the same fungus (holomorph).

Recently, considerable progress has been made towards establishing the connection between a particular anamorph and its teleomorph by comparing DNA sequences, but there are still many fungi for which this remains to be done. For fungi that have lost the capability to reproduce sexually, there is no teleomorph, and the anamorph alone constitutes the holomorph.

Septate Hyphae

The vast majority of ascomycetes (with the yeasts representing the major exception) produce a mycelium consisting of a system of hyphae. In this respect they are not unlike members of the Zygomycota. Only closer examination of an individual hypha reveals a significant difference, because in the Ascomycota the hyphae are regularly septate, while in the Zygomycota the hyphae lack septa and thus are coenocytic.

The presence of septa confers a number of advantages. First, the septa give the hypha a degree of physical rigidity. Second, because the septa divide the hypha into cell-like compartments, damage (that is, the loss of cytoplasm) is confined largely to just the affected compartments and not to the entire mycelium as would be the case in the Zygomycota. In an evolutionary sense, this apparently allows a septate mycelium to "take more chances" than one that lacks septa. For example, ascomycetes and members of the Basidiomycota (which also have septate hyphae) can survive and produce fruiting bodies under a wider range of environmental conditions than members of the Zygomycota.

Each septum in a hypha is perforated by a minute central pore, which is so small as to be difficult to observe even under a microscope. As such, the cytoplasm of one cell-like compartment is continuous with that of its neighbors. Some of the smaller cell organelles such as mitochondria can pass through the pores from one compartment to the next, but the larger nuclei do not. The usual situation is for each compartment to contain a single nucleus, although this is not always the case (as will be described in Chapter 4).

Asexual Spores

The production of asexual spores is the dominant form of reproduction for the ascomycetes as a group. These asexual spores are invariably called conidia (singular: conidium), although use of the terms "conidiospore" or "mitospore" is not incorrect. Conidia are the result of the process of mitosis and thus are genetically identical to the mycelium from which they are ultimately derived. Most have just one nucleus. They are generally formed from specialized cells called conidiogenous cells that are produced on the ends of specialized hyphae called conidiophores. (One should note that the words *conidiophore* and *conidiospore* are very similar and thus easily confused. This is a good reason for using the word *conidium* for an asexual spore in the ascomycetes.)

The fungi that produce conidia can be divided into two main groups, the hyphomycetes and the coelomycetes. The hyphomycetes produce their conidia directly from vegetative hyphae or on conidiophores that arise singly from the mycelium, whereas the coelomycetes produce conidia on tightly packed masses of conidiophores located within more or less flask-shaped fruiting bodies called pycnidia (singular: pycnidium). To say that these two groups include fungi that exhibit an extraordinary diverse range of shapes, forms, and colors would be an understatement.

Classification of Ascomycota

Three subphyla are generally recognized for the phylum Ascomycota. The first and by far the largest of these is the subphylum Pezizomycotina, which is essentially made up of ascomycetes that are filamentous and characterized by the production of either microscopic reproductive structures with asexual spores or fruiting bodies (ascocarps) with asci and ascospores. In many instances, as will be described below, different forms of the same species of fungus can embody both of these characteristics.

The second subphylum (the Saccharomycotina) includes those ascomycetes that make up the "true" yeasts. These are generally single-celled or consist of no more than short chains of cells and reproduce vegetatively by budding rather than by the production of asexual spores.

The third subphylum (the Taphrinomycotina) is a relatively small but heterogeneous assemblage of fungi that was only recognized as a distinct group after data on DNA sequences became available from molecular analyses. The Taphrinomycotina is considered the most primitive of the three subphyla. Taxa assigned to this group range from free-living saprotrophic, sometimes yeastlike forms, to fungi that are obligate parasites of plants.

In this chapter, only the Saccharomycotina and those members of the Pezizomycotina for which the anamorph represents the predominant stage encountered in nature will be discussed. In some instances, as noted above, the anamorph appears to be the only stage in the life cycle, with the capability of reproducing sexually having been lost. In other instances, the teleomorph is also known for a particular taxon, but this stage will be considered only in the context of the information presented in Chapter 4.

Subphylum Pezizomycotina

As a group, the anamorphic members of the Pezizomycotina are ubiquitous, occurring in virtually every imaginable type of environmental situation. They are especially abundant in soils and are invariably found in association with all kinds of organic matter. Most are saprotrophs, but some are parasites of plants and a few examples produce infections in animals (including man). In a taxonomic sense, this is a vast assemblage of fungi. The total number of species has been estimated at 20,000 (Kiffer and Morelet 2000), but this is undoubtedly a conservative figure.

Some indication of the incredible diversity of microfungi associated with particular substrates in nature was provided by a study carried out by Gerald Bills and Jon Polishook, two microbiologists who worked at the Merck Research Laboratories in New Jersey. They employed a modified version of a particle-filtration technique used in soil ecology to examine the microfungi associated with samples of leaf litter collected in a lowland rain forest in Costa Rica (Bills and Polishook 1994). In brief, their technique involved breaking up the samples into very fine particles, washing these repeatedly to remove any fungal spores that might be present, and then spreading small amounts of this material onto the surface of agar in a series of culture plates. Presumably, most of the fungi appearing in these plates would have developed from fragments of hyphae actually present on or within dead leaves. Most of the other species not directly associated with the dead leaves and thus not part of this ecological assemblage would have been eliminated by the washing. The fungi that developed in the plates were identified to the extent possible. Numbers of species per sample ranged from 78 to 134. These are amazing totals from what were relatively small samples from just one part of the total forest habitat.

Because these fungi are so common and widespread, it is not surprising that their spores are exceedingly common in the air, both indoors and in nature. Given the opportunity (which is often difficult to prevent from happening), their spores settle upon and then rapidly produce vegetative mycelia on all types of foodstuffs. Stored cereals, fruits, dried meats, bread, and jams are all susceptible, and the annual worldwide loss of food to these fungi is almost incalculable. However, as will be described below, certain anamorphic members of the Pezizomycotina are also among the more important of all fungi in terms of their positive aspects.

Aspergillus and Penicillium

In the absence of molecular data that have become available only recently, classification of the anamorphic fungi was based almost entirely upon features that could be directly observed with the use of a microscope or noted for actively growing mycelia in laboratory culture. Among the more important considerations were the overall morphology of the conidiophores upon which conidia are produced and the color of the conidia as viewed in mass when numerous conidiophores are present on an actively growing mycelium. Two very large and economically important genera—*Aspergillus* and *Penicillium*—provide some indication of the extent of the variation that exists for different anamorphic taxa.

In *Aspergillus*, a genus that consists of more than 200 species, the conidiophore is erect and unbranched, with an apex that is swollen to form a globular vesicle. Numerous secondary conidiogenous cells called phialides radiate outward from the vesicle in all directions (PLATE 23). The phialides are special cells from which individual conidia are budded off. If undisturbed, the conidia form a short chain from the apex of a phialide; however, they are easily detached and carried away on air currents. The color of the mature conidia in mass varies from species to species. Some of the more commonly encountered colors include white, yellow, blue, and black.

In *Penicillium*, an even larger genus of at least 250 species, the conidiophore is branched above the middle, and each of the more or less parallel branches has a phialide at its tip (PLATE 24). Each phialide buds off a chain of individual conidia in the same manner as described for *Aspergillus*. The mature conidia of *Penicillium* are usually yellow, blue, or blue-green in color.

Members of both genera are exceedingly common as saprotrophs, and *Penicillium* is perhaps the most diverse of all filamentous fungi in soils. Interestingly, species of *Aspergillus* tend to be more abundant in warmer regions of the world, while species of *Penicillium* are generally more characteristic of cooler regions.

Both genera include species that are important to humans. *Penicillium chrysogenum* is the source of penicillin, an antibiotic that has saved countless lives. Other species of *Penicillium* are used to make certain kinds of cheese. The former include *P. roqueforti* for blue-veined cheeses and *P. camemberti* for white mold or Brie (named after a region in France) cheeses. Interestingly, when these two species occur on substrates other than milk or milk products, they have a negative rather than a positive value. In fact, *P. roqueforti* is a common contaminant on corn and other types of agricultural silage, where it sometimes represents a serious problem for farmers (O'Brien et al. 2006). The most negative aspects of a fungus such as *P. roqueforti* are the result of mycotoxins produced under some circumstances.

Mycotoxins are chemicals that evoke a toxic response when ingested by humans and other animals. More than 200 different mycotoxins are now known, and this number continues to increase as additional fungi are subjected to study. Mycotoxins typically cause problems when they are ingested, often in low concentrations, over a relatively period of time. For humans, most mycotoxins are derived from fungi that have contaminated improperly stored food. Numerous species of *Penicillium* can contaminate food and thus have considerable negative importance as a result of the mycotoxins they produce.

The same is true for many species of *Aspergillus*, which are often serious contaminants of starchy food, including various cereal grains. *Aspergillus flavus*, which can produce the extremely deadly and potentially carcinogenic mycotoxins known as aflatoxins under certain special conditions, is particularly noteworthy. Aflatoxins were first identified in 1960, when an estimated 100,000 young turkeys on poultry farms in England died from what was first referred to as turkey X disease, but then was linked to the consumption by the turkeys of peanut meal containing lethal amounts of unknown toxins. The toxins were found to have been produced by *A. flavus*, which had contaminated the peanut meal. The word *aflatoxin* is actually derived from the name of the fungus (with "afla" representing the first part of *A. flavus*) from which this type of mycotoxin was first reported (Kendrick 2000).

Clearly, mycotoxins produced by fungi, including species of *Aspergillus* and *Penicillium* as well as various other taxa, have posed health hazards for humans as long as the storage of food (especially cereal grain) has been a common practice. At first, both the amount stored and the length of time it remained in storage were limited, so mycotoxins may not have been a major problem. As soon as grain was stored on a larger scale in specially constructed storage facilities (granaries), the likelihood of contamination by fungi (and thus the potential for mycotoxins) would have increased. Even in ancient societies, efforts would have been made to limit the occurrence of moldy grain by keeping the conditions for storage relatively dry and providing adequate ventilation. Not surprisingly, this same approach is just as effective today.

Some species of *Aspergillus* can affect humans more directly. For example, *A. fumigatus* is an opportunistic pathogen of humans and other animals. In humans, it is responsible for aspergillosis, a pulmonary disorder that is one of the most common causes of death from hospital-acquired infections. On a more positive note, *A. oryzae*, which is very similar to *A. flavus* and is considered by some mycologists to be the same species (Seifert, personal communication), is used in the production of saké, an alcoholic beverage made from rice. *Aspergillus niger* is the major industrial source of citric acid, accounting for 99 percent of global production. The latter species, although generally recognized as being a "safe" fungus, is capable of producing a type of potentially carcinogenic mycotoxin known as ochratoxin A.

When the teleomorph stage has been identified for species of *Penicillium* and *Aspergillus*, the fungi involved represent a number of different genera. These include *Emericella* and *Eurotium* for *Aspergillus*, and *Eupenicillium* and *Talaromyces* for *Penicillium*. All of these genera are members of the order Eurotiales, a group that will be described in Chapter 4. Numerous other genera of anamorphic fungi have been described. Among the more commonly encountered of these are *Fusarium*, *Gliocladium*, *Trichoderma*, and *Verticillium*. Their teleomorphs are found in various orders of the Pezizomycotina other than the Eurotiales.

Identifying Terrestrial Mitosporic Ascomycetes

Because the mitosporic ascomycetes are so taxonomically and morphological diverse, they do not lend themselves to being easily classified or identified. This situation has posed a major problem for anyone who needs to "put a name" on some fungus of particular concern (for example, as a pathogen or food contaminant).

The first useful system of classification was developed by the Italian mycologist Pier Andrea Saccardo. Over a period of 35 years, Saccardo

(1882–1931) attempted to compile a comprehensive index to all known fungi (Totter 1972). As a result of this effort, he was forced to devise a classification system for the mitosporic ascomycetes so that he could include them in his index. The system was artificial rather than taxonomic and was based upon the overall morphology and color of the spores and the structures upon which the spores were produced.

Although no longer valid, the Saccardo system represents an effective way to arrive at an identification of an unknown mitosporic ascomycete. This system forms the basis of the keys provided in *Illustrated Guide of Imperfect Fungi* (1998), written by Harold Barnett, a plant pathologist at West Virginia University, and Barry Hunter, a mycologist at California University of Pennsylvania. Since the book first appeared, many mycologists have considered it to be the first thing to consult when confronted with the task of identifying an unknown fungus. Moreover, the line drawings (which have been supplemented with some photographs in later editions) also provide a visual survey of the myriad forms that make up this assemblage of fungi.

Yeasts

When I took my first graduate course in mycology at Virginia Tech, a lot of information on fungi was totally new. Some of it was the type of stuff that students learn for a test or exam but then forget until it is encountered again in another course or reacquired in some other context (for example, writing or reviewing manuscripts or developing some new aspect of a research program). However, as anyone who has been a student can attest, there are some tidbits of knowledge that once delivered by the professor are never forgotten. One of the things I remember most from that mycology class, which was taught by Orson Miller, was the simple statement that "yeasts are everywhere."

Obviously, most people have heard of yeasts in the human context (a subject to be discussed in Chapter 11), but the point Miller was trying to make was not limited to just the yeasts used in baking bread and fermenting alcoholic beverages. He was talking about all yeasts.

To begin with, the term "yeast" merely denotes a certain type of growth form in the fungi and not a particular taxonomic group. In fact, two different phyla (Ascomycota and Basidiomycota) include taxa that are considered as yeasts. Nevertheless, the best-known yeasts are members of the Ascomycota.

By definition, yeasts are unicellular fungi that reproduce by budding, a process in which a daughter cell (or bud) is formed directly from the original (or parent) cell. The nucleus of the parent cell splits into two nuclei and one of these migrates into the bud. The bud continues to grow until it eventually separates from the parent cell, forming a new cell. In some yeasts, the first bud produced may in turn give rise to a second bud before the former has separated completely from the original cell. The result is a short chain of cells, which can be referred to as a pseudomycelium, but the basic form of the yeast is still unicellular.

Budding is a method of asexual reproduction. While some yeasts appear to have lost the capability for sexual reproduction, many others reproduce sexually, although sometimes only rarely. For those yeasts that are members of the Ascomycota, sexual reproduction usually involves little more than the direct transformation of a diploid vegetative cell into an ascus. Meiosis occurs within the cell to produce four ascospores. These are liberated when the wall of the ascus breaks down. Ascospores are far more resistant structures than vegetative cells and can be regarded as the resistant stage in the life cycle.

Ultimately, an ascospore germinates to give rise to a haploid vegetative cell. The haploid vegetative cell is similar in appearance to, albeit slightly smaller than, the diploid vegetative cell. A given population of some species of yeasts may consist of diploid cells, haploid cells, or a mixture of the two types. When haploid cells of different mating types (a situation similar to what was described for *Rhizopus stolonifer* earlier in this chapter) encounter one another, they fuse and restore the diploid condition. As a result, the yeast cells in the example mentioned above are most likely to be predominately diploid.

Ascomycetes—The Best-known Yeasts

Saccharomyces cerevisiae (the genus name literally means "sugar fungus") is the best-known yeast (PLATE 25). The main reason for this is that *S. cerevisiae* has been used by humans for thousands of years to ferment alcoholic beverages and bake bread. Because of its economic importance, the species has received considerable study. Moreover, its simple structure has resulted in *S. cerevisiae* being used as a model organism in basic scientific research relating to the fundamental biology of the cell. Consequently, there is some justification for regarding this unicellular fungus as one of the most intensively studied of all living organisms.

The single most important feature of *Saccharomyces cerevisiae* from a human point of view is the ability of this yeast to produce ethyl alcohol and carbon dioxide from sugar under anaerobic conditions. The yeast cell metabolizes the sugar molecules. If oxygen is present (aerobic conditions), carbon dioxide and water are produced. In the absence of oxygen or in the presence of a very high concentration of sugar, the cell switches to a different metabolic pathway and fermentation occurs, which yields carbon dioxide and ethyl alcohol.

Basidiomycete Yeasts

In 1680, Anton van Leeuwenhoek, the Dutch naturalist who had invented the microscope, was the first person to use this instrument to observe the cells of yeasts. He considered them to be globular structures and not living organisms. Almost two centuries later, the French microbiologist Louis Pasteur demonstrated that the production of alcoholic beverages through the process of fermentation was carried out by living yeasts and thus was not just a reaction to some chemical catalyst.

The knowledge that some yeasts are members of the Basidiomycota and not the Ascomycota is a relatively recent development in mycology. Although suggested by some mycologists as early as the 1920s, it was not until the mid-1960s, when a teliospore stage was described in the cycle of the yeast *Rhodosporidium toruloides*, that yeasts were clearly known to belong to the Basidiomycota (Banno 1967). As will be described in Chapter 9, teliospores are produced only in a certain group in this phylum. It is now recognized that yeasts occur in all three of the major subphyla of the Basidiomycota. For example, *R. toruloides* belongs to the subphylum made up of those fungi commonly referred to as rusts.

Some of the yeasts assigned to the Basidiomycota are known to have both anamorph and teleomorph states, with the former almost always the better known. Some of the more common and widely distributed examples are found in the anamorph genus *Rhodotorula*. These are among the more distinctive yeasts, since they form colonies that are pink to red in color. Species of *Rhodotorula* are now known to have telcomorphs in the genus *Rhodosporidium*.

Other Fungi That Produce Yeastlike Cells

A few fungi that do not fit the definition of yeasts sometimes produce yeastlike cells under special conditions. Such is the case for certain species of *Mucor* in the Zygomycota. These fungi are ordinarily filamentous, but they are capable of forming yeastlike cells that can produce buds. Moreover, these same yeastlike cells can ferment sugars into alcohol. This would appear to be a remarkable example of convergent evolution involving two very different groups of fungi.

Yeasts in Nature

In nature, the cells of *Saccharomyces cerevisiae* commonly occur on the surfaces of ripe fruits, including grapes. Like most species of yeasts, *S. cerevisiae* is a saprotroph. Presumably, the accidental discovery that the sugary juice of grapes could be converted into an alcoholic beverage took place only because the yeasts involved just happened to be present on the skin of the grapes.

Many different kinds of yeasts are associated with plant surfaces, where they exploit various substances exuded by the plant. For example, yeasts are normal inhabitants of leaves and stems. They are especially abundant on fruits and the nectar-producing parts of flowers. Various species of yeasts are commonly found in soil, including soils of every type of ecosystem examined to date. A few species are even known from soils of the Antarctic. Moreover, as will be discussed in more detail in Chapter 10, certain species of yeasts are found on the body surfaces or within the digestive tracts of insects.

Yeasts are now known to be the dominant fungi in the open oceans, although both overall density (typically below 10 cells per quart [0.95 liter]) and the number of species present are low. This is not surprising, since seawater typically contains low levels of the organic molecules that yeasts need to meet their energy needs. Surprisingly, the assemblage of species present in the ocean is not unlike those found in freshwater or terrestrial habitats, and only a very few yeasts appear to be restricted to the marine environment.

Yeasts in the phylum Basidiomycota are generally more common than those that belong to the Ascomycota in open oceans, but the reverse is true for coastal areas. For example, yeasts in the Ascomycota are extremely diverse and often abundant in polluted estuaries and in the waters of subtropical mangrove forests. In these habitats, yeasts appear to represent an important food source for some small marine invertebrates.

Numbers of yeast cells in sediments of intertidal estuaries and similar habitats are many times higher than those of the open water. Representatives from both the Ascomycota and Basidiomycota commonly occur in freshwater ponds and lakes, and the assemblages of species present have been reported to change from one season to next. In some instances (for example, recreational lakes), these changes have been linked to human activities. The total number of yeast cells increases as the level of pollution increases. Unpolluted lakes may have fewer than 100 cells per quart (0.95 liter), but the number in polluted lakes is several orders of magnitude higher. Since some species of yeast characteristically occur only when the water becomes polluted, they have been used as indicators of organic pollution (Nagahama 2006).

In short, Miller was right—yeasts are indeed everywhere.

4
A Diversity of Form and Function

Sexual Reproduction in Phylum Ascomycota 67
Types of Ascomycete Fruiting Bodies 69
Ascus-producing Terrestrial Microfungi 70
Order Capnodiales 71
Order Clavicipitales 71
Claviceps and Ergot of Grasses 72
Cordyceps and Parasites of Insects 72
Order Cyttariales 73
Order Diaporthales 74
Order Erysiphales 74
Order Eurotiales 75
Order Helotiales 75
Botrytis 76

Jelly Clubs and Other Helotiales in Nature 76
Earth Tongues 77
Order Hypocreales 77
Lobster Mushroom Fungus 77
Tree Pathogens 78
Order Laboulbeniales 78
Order Ophiostomatales 79
Dutch Elm Disease 80
Blue-Stain of Wood 81
Order Pleosporales 82
Black Knot of Cherry 82
Apple Scab 82
Order Sordariales 83
Fungi on Burnt Plant Matter 83
Dung-loving Ascomycete Fungi 84
Fabric- and Paper-loving Fungi 84

The terrestrial microfungi considered in the preceding chapter share two features. First, their spores are produced in fruiting structures too small to be detected easily in nature. In many instances, this fruiting structure is nothing more than a single specialized hypha. As such, the use of the term "fruiting body" would not be regarded as appropriate.

Second, the spores produced are mostly asexual spores, often referred to as conidia.

In this chapter, those members of the phylum Ascomycota characterized by the production of a fruiting body (called an ascocarp or ascoma) containing sexual spores (ascospores) will be considered. This is a large and heterogeneous assemblage of fungi, and their fruiting bodies exhibit an amazing diversity of form and function. However, there is an underlying similarity in how they develop.

Sexual Reproduction in Phylum Ascomycota

Sexual reproduction involving the development of an ascus and ascospores is a highly distinctive process in the Ascomycota. When an ascospore or a conidium germinates to produce an initial hypha that grows and proliferates to become a mycelium, both the original nucleus in the spore and all of the nuclei derived from it through a series of mitotic divisions are haploid and of one type. In some cases, the ability to reproduce sexually has been lost, and the fungus exists only in the haploid state. In other cases, the fungus is capable of sexual reproduction, and when circumstances allow, the mycelium may need to come into contact with another compatible mycelium of the same species. This event can be observed for fungi grown in laboratory cultures but is impossible to observe directly in the field.

For those species with mycelia that tend to be widely scattered in nature because the substrates they occupy are uncommon, contact between compatible mycelia is a rare event. In any case, when two mycelia of the same species do meet, some of their hyphae (or specialized cells produced by those hyphae) fuse with one another. This can take place in a number of ways, and various specialized structures are sometimes involved. The net effect is that a new system of hyphae is formed, and the individual hyphae (called ascogenous hyphae) that make up this new system contain equal numbers of nuclei from each of the original mycelia. The nuclei from the two different sources pair and then divide synchronously so that the 1:1 relationship is maintained.

This phenomenon is not unique to the Ascomycota, since it also occurs in the Basidiomycota. For this reason the two phyla are sometimes treated as a single large taxon (the subkingdom Dikarya) in the kingdom Fungi. The name of the subkingdom is derived from the fact that each of the pairs of nuclei referred to above represents a dikaryon (essentially meaning two types of nuclei), and the hyphae in which they occur are referred to as being dikaryotic. By extension, the hyphae making up each of the two original mycelia are monokaryotic, since they contain only a single type of nucleus.

In the Ascomycota, the dikaryotic condition is found only in the ascogenous hyphae. Although the system of ascogenous hyphae can become quite extensive in some large fruiting bodies, these hyphae actually persist for a relatively short period of time in the overall life cycle. In contrast, the dikaryotic condition is the dominant one in the Basidiomycota.

The ascogenous hyphae described above consist of linear series of cell-like compartments, each with two nuclei (the dikaryon described above). In the actual formation of an ascus, the two nuclei within a terminal cell (which is to become the ascus) of an ascogenous hypha fuse to form a diploid nucleus. This nucleus immediately undergoes meiosis so that four haploid nuclei are formed. In most of the Ascomycota, these then divide mitotically to yield a total of eight nuclei, and each of these develops into an ascospore. The end result is an ascus, usually with eight ascospores. In some cases, multiple mitotic divisions occur, resulting in asci with hundreds or even thousands of ascospores.

As indicated in Chapter 3, the ascus is the single defining characteristic of the phylum Ascomycota. In situations where multiple asci are produced, the individual asci are often dispersed among other sterile hyphae called paraphyses (singular: paraphysis). Paraphyses can be about the same length as or longer than the adjacent asci. When the ascospores are mature, they are liberated and subsequently dispersed from the ascus in various ways. Some asci develop a terminal pore through which the ascospores are forcibly discharged, while others break open on contact with water, simply disintegrate with the passage of time, or are disrupted through the actions of some animal. When an ascospore reaches a suitable substrate, it germinates to form new hyphae; the sexual life cycle has come full circle.

In many ascomycetes, numerous asci occur together in small groups (fascicles) or form a distinct layer (called the hymenium) within a clearly recognizable fruiting body. In other members of the phylum no fruiting body is formed, and asci are produced singly, in low numbers or are found as a layer within the tissues (that is, beneath the epidermis) of some host plant.

Types of Ascomycete Fruiting Bodies

The fruiting body can take one of any number of different morphological expressions, depending on the species involved. However, there are essentially three basic types. In the first type, the asci occur on the inside of a more or less spherical fruiting body called a cleistothecium (plural: cleistothecia). In the second type, the asci occur within a flask-shaped fruiting body, referred to as a perithecium (plural: perithecia). In the third type, the fruiting body, known as an apothecium (plural: apothecia), is more or less disk- or cup-shaped, and the asci occur in a layer (hymenium) over the upper surface. In the very largest examples of apothecia, the number of asci present can exceed several million. Of the three basic types of fruiting bodies, the apothecium is probably the most familiar to non-mycologists and provides the basis of a common name, cup fungi, sometimes applied to the ascomycetes as a group.

It would be a mistake to think that all of the fruiting bodies produced by ascomycetes are morphologically simple. Some occur as composite structures, with, for example, many perithecia grouped together. Moreover, fruiting bodies may be stalked or sessile, solitary or clustered, and they can range from almost microscopic to structures that are large enough to be easily noticed. Some are club-shaped, while others cushion-shaped, coral-like, or even golfball-shaped (PLATE 27). Fruiting bodies can be fleshy or carbonaceous (that is, like charcoal), leathery, rubbery, gelatinous, slimy, or even powdery. They come in a multitude of colors, including red, orange, yellow, brown, black, and (more rarely) green or blue, but brown or black are the colors most commonly encountered.

The extraordinary range of morphological diversity expressed by members of the Ascomycota has been variously interpreted by different mycologists and is reflected in the different taxonomic classifications that have been proposed for this group. Although the new body of data available from molecular phylogenetics has eliminated much of the taxonomic uncertainty as to the placement of particular taxa, some questions still remain, and there is as yet no single totally comprehensive and universally accepted classification scheme. Nevertheless, some of the more traditional approaches based on the overall morphology of the fruiting body still work well enough for the purposes of the present discussion.

As mentioned above, three basic types of fruiting bodies can be recognized, and those ascomycetes that characteristically produce one of these types can be assigned to an assemblage and given a name. For example, those fungi that produce cleistothecia have been referred to as plectomycetes by mycologists. (This is yet another example of a term that was once used to denote a taxonomic class.) In the same way, those fungi in which the fruiting body is a perithecium have been called pyrenomycetes, and the ones that produce apothecia have been known as discomycetes. Sometimes, appearances can be deceiving. In some ascomycetes (a group recognized as the loculoascomycetes), the fruiting bodies often closely resemble perithecia but have a different type of development and asci that are structurally different than those produced by the taxa assigned to the plectomycetes, pyrenomycetes, and discomycetes. This type of perithecium-like fruiting body is called a pseudothecium.

Although the groups delimited above are still being used by some mycologists, they are not supported by the data obtained from molecular-based studies. Even in the absence of molecular data, mycologists who work with these fungi have long recognized certain core groups usually considered to represent taxonomic orders. More than several dozen orders have been listed in some recent comprehensive treatments of the classification of the ascomycetes, but some of these are of limited occurrence or have little ecological or economic importance.

Ascus-producing Terrestrial Microfungi

Fungi likely to be encountered in nature or those of considerable economic or ecological importance are included in the twelve orders considered in this chapter. These are the Capnodiales, Clavicipitales, Cyttariales, Diaporthales, Erysiphales, Eurotiales, Helotiales, Hypocreales, Laboulbeniales, Ophiostomatales, Pleosporales, and Sordariales. If the more traditional system of classification described above, which is based solely on morphology, is applied to these orders, two

(Erysiphales and Eurotiales) would be assigned to the plectomycetes, two (Cyttariales and Helotiales) to the discomycetes, two (Capnodiales and Pleosporales) to the loculoascomycetes, and the others to the pyrenomycetes. Each of these includes at least some taxa common, interesting, or important enough to warrant discussion. For the most part, the fruiting bodies produced by members of these orders are small and relatively inconspicuous, although a few examples are of sufficient size or colorful enough to be easily noticed. Two other orders, the Xylariales and the Pezizales, are made up of species for which the fruiting bodies are often large enough to be fairly conspicuous, and these will be covered in Chapter 5.

Order Capnodiales

In southern beech (*Nothofagus*) forests of some parts of the South Island of New Zealand, the trunks, branches, and leaves of many trees appear black because they are almost completely covered by the mycelia of sooty molds. These fungi are representatives of the order Capnodiales and occur on the sugary wastes (honeydew) secreted by scale insects. Careful examination of the trunk of one of the trees with a thick coating of mycelium present will usually reveal some of the glistening droplets of honeydew.

Sooty molds also can be found on leaves of many other plants, including common ornamentals such as azaleas, gardenias, and crape myrtles. In each instance, the presence of the fungus can be related to the honeydew produced by insects (aphids and whiteflies in addition to scales) that suck sap from the host plant. The fungus itself apparently does little harm to the plant other than blocking sunlight for portions of the plant covered by the mycelium. As such, sooty molds represent little more than a cosmetic problem for humans who view affected plants.

Order Clavicipitales

As traditionally recognized, the order Clavicipitales consists of fewer than 300 species of fungi that either are pathogens or endophytes of various grasses and other plants or occur as parasites of insects or the fruiting bodies of other fungi. Some mycologists do not recognize these fungi as constituting a separate order. Instead, they are assigned to a family (Clavicipitaceae) in the Hypocreales. Whatever the taxonomic approach used, there is little question that some rather distinctive species of fungi are involved. Many of the better-known examples belong to just two genera—*Claviceps* and *Cordyceps*. The former includes species associated with grasses, while members of the latter are associated with insects and fungi.

Claviceps and Ergot of Grasses

The best-known species of *Claviceps* is *C. purpurea*, which causes ergot of grasses. This fungus is widely distributed in temperate regions of the world, where it has an exceptionally wide host range (more than 400 different species of grasses) for a pathogenic fungi. One of the more important host plants is rye grass (*Secale cereale*).

The ergot is a mycelial mass that the fungus produces when it infects the developing ovary in a flower of the host. When it first appears, the mycelial mass is white and cottony, and at this stage gives rise to numerous short conidiophores and conidia. The mycelial mass secrets a sugary, nectarlike substance that attracts insects. The insects come into contact with the mass, pick up conidia, and then disperse them to other potential host plants. Eventually, the mycelial mass hardens into a purplish black sclerotium, the structure known as the ergot. The latter contains high concentrations of several deadly alkaloids.

If the host grass is being grown as a crop, the ergots are likely to be gathered along with the grain. If consumed by humans, the alkaloids in the ergots can cause a form of poisoning known as ergotism. During the Middle Ages, when rye bread was common in the diet of people in some parts of Europe, outbreaks of ergotism were common, oftentimes leading to many deaths. In humans, ergotism produces a burning sensation in the limbs, and thus became known as St. Anthony's Fire in the affected individual, who also tended to suffer hallucinations.

With improved methods of cleaning and milling grain and a switch away from rye bread to wheat bread, ergotism is now rare in humans. It should be noted that the alkaloids produced by *Claviceps purpurea* have received some use in medicine. For example, if properly administered, they can restrict hemorrhaging following childbirth.

Cordyceps and Parasites of Insects

Species of *Cordyceps* produce some of the most bizarre fruiting bodies found in the fungi. In this genus, numerous perithecia are embedded in the upper portion (which often takes the form of a distinct "head") of an elongated, clublike stroma that arises directly from the body of the infected host. In most instances, the host is an insect, but in other cases the host is another fungus. Such is the case for *C. ophioglossoides*, which is parasitic on the subterranean fruiting bodies of false truffles in the genus *Elaphomyces*.

Cordyceps militaris (orange club) is a commonly encountered parasite of insects. The bright orange to orange-red fruiting bodies of this fungus develop from the buried pupae of moths. Most of the time a single fruiting body is present, but it is not unusual to find several attached to the same pupa. Although not large (most fruiting bodies are no more than 1¼ to 2 inches [3 to 5 cm] tall), the bright color of *C. militaris* makes it easy to spot against the darker background of the dead leaves on the forest floor where the fungus usually occurs.

Cordyceps sinensis (vegetable caterpillar fungus) occurs in Asia and has long been used in traditional Chinese medicine. This fungus is parasitic on the larvae (caterpillars) of moths, and its fruiting bodies may reach lengths of 4 inches (10 cm) or more. The combination of moth and fungus (known as a "vegetable caterpillar") is collected and used as a treatment for a number of human maladies. More recently, some Chinese distance runners have attributed improvements in their performances to the use of vegetable caterpillars.

A similar species (*Cordyceps robertsii*) occurs in New Zealand (PLATE 26).

Order Cyttariales

Some of the other most distinctive fruiting bodies found in the ascomycetes are produced by members of the order Cyttariales. Species of *Cyttaria* occur as parasites on certain species of southern beech in southern South America, Australia, and New Zealand. The fruiting body is a globose, fleshy stroma within which numerous apothecia are embedded (PLATE 27). These structures, which generally arise from areas of deformed tissue (galls) on the branches of an infected tree, can reach the size of a golf ball.

Some species of *Cyttaria* are edible. For example, *C. gunnii* (beech strawberry) was a traditional food item for the aborigines in those parts of Australia where southern beech occurred, and *C. espinosae* is still commonly eaten in southern South America.

Order Diaporthales

The order Diaporthales includes approximately 500 species of fungi that are mostly saprotrophs or parasites. Several examples are exceedingly important as pathogens of certain species of trees and are the most likely to be known to the average person. These tree pathogens typically produce perithecia that are embedded in the tissue of the host tree, with just the long neck of the perithecium protruding beyond the surface of the bark. In many instances, pycnidia of the anamorph are more apparent than the perithecia on the bark of the infected tree.

The most widely known member of the Diaporthales is *Cryphonectria parasitica* (formerly known as *Endothia parasitica*), the infamous fungus that caused the virtual disappearance of American chestnut from the forests of eastern North America. The devastation of American chestnut by *C. parasitica* will be described in more detail in Chapter 11.

Another member of the order is *Discula destructiva*, which infects dogwood and causes dogwood anthracnose. Apparently introduced into North America in the 1970s, this disease has had a devastating effect upon populations of flowering dogwood (*Cornus florida*) throughout eastern North America.

Species of *Diaporthe* and its anamorph *Phomopsis* are the causal agents of a number of other plant diseases, some of which are economically important.

Order Erysiphales

The members of the order Erysiphales are obligate parasites of plants. The 500 or so species of this order form a clearly defined group of fungi commonly called powdery mildews because the infected portions of the host have a white, powdery appearance as a result of the presence of a superficial mycelium.

Unlike most other fungi that infect plants, the mycelium of a powdery mildew does not proliferate throughout the tissues of the host. Instead, the mycelium forms an incomplete and superficial covering that extends over the external surface (epidermis) of a leaf. This mycelium produces specialized hyphae that penetrate cells of the epidermis and develop into special absorptive structures called haustoria through which the fungus derives its nourishment from the plant. During the summer months, the mycelium gives rise to numerous conidiophores, and conidia produced by the latter are dispersed to new host plants by the wind. In the fall, cleistothecia appear on the mycelium. When mature, these are usually dark brown or black and appear as small dark dots against the white background of the mycelium when the latter is viewed with a hand lens. Each cleistothecium contains one to several asci, from which ascospores are eventually discharged when the wall of the cleistothecium ruptures.

The cleistothecia of powdery mildews are distinctive because of the presence of special hyphae (known as appendages) that radiate from the outer wall (PLATE 28). The taxonomic concepts used to delimit different genera of powdery mildews have been based largely upon morphological features of these appendages along with the number of asci in the cleistothecium itself. For example, members of the genus *Uncinula* are characterized by appendages with curled tips, while members of the genus *Phyllactinia* have needle-shaped appendages with a bulbous base.

Order Eurotiales

The order Eurotiales, as was mentioned in Chapter 3, includes the teleomorphic stages of such important genera of anamorphic fungi as *Aspergillus* and *Penicillium*. It is for this fact alone that the order warrants inclusion in the present discussion. About 140 species have been assigned to this order, and in most of those for which the teleomorph is known, the fruiting body produced is a cleistothecium. Numerous other species are known only as anamorphs. Because of the dominance of the anamorph in this group of fungi, it is the asexual form that receives the bulk of the attention.

Order Helotiales

Although most members of the taxonomic orders in this chapter produce small, inconspicuous fruiting bodies, exceptions do exist, and several members of the order Helotiales (also listed as the Leotiales in some treatments of the fungi) represent good examples.

On the basis of the type of fruiting body (an apothecium) produced, the Helotiales would be considered as discomycetes, which is also the case for one of the two orders (Pezizales) to be considered in Chapter 5. Although the fruiting bodies of some members of the two orders can be superficially similar, they differ in one very important respect. In the Leotiales, the asci are said to be inoperculate. What this simply means is that the mature ascus does not have a terminal pore with an associated lid (called an operculum). In the Pezizales, the asci are operculate, since they do have an operculum. Mycologists often use the terms "inoperculate discomycetes" and "operculate discomycetes" to refer to the fungi in the two orders.

The Helotiales is large order of about 2000 species. Many of the taxa assigned to this order have well-known anamorphs that are important as causal agents of various plant diseases. Among these are species belonging to the anamorph genera *Monilia* and *Botrytis*.

Botrytis

Botrytis cinerea has been recorded as a pathogen on more than 200 different species of plants. It is economically important for soft fruits such as strawberries, where it causes the disease known as gray mold. The affected strawberries are not edible and have to be discarded. *Botrytis cinerea* also infects overripe grapes, but in some wine-producing areas of Europe such grapes are prized and not discarded. In fact, the grapes are not picked but are left on the vine on purpose, so that they can be subject to infection by *B. cinerea* (or "noble rot" as it is called). Infected grapes crack open, and the fungus lives as a saprotroph on the juice. This causes the sugar content of the grapes to increase, and at the proper time, they are picked and made into wine. The resulting wine is much sweeter and richer than ordinary wines, and commands higher prices.

Jelly Clubs and Other Helotiales in Nature

Some of the other members of the order produce fruiting bodies that are often encountered in nature. For example, clusters of the small bright lemon-yellow fruiting bodies of *Bisporella citrina* (yellow fairy cup) commonly occur on decorticated wood.

Members of the genus *Chlorociboria*, which are found throughout the world, are characterized by a mycelium that stains the wood in which it occurs a bluish-green color. The fungus (commonly referred to as green stain of wood) can be identified from this feature alone. The fruiting bodies, which are small, often no larger about $3/8$ inch (1.0 cm) across, and infrequently encountered, are also blue green in color (PLATE 29).

The fruiting bodies of *Leotia lubrica* (jelly club) are stalked, jellylike and have a somewhat contorted convex upper surface (head) upon which the hymenium occurs. The yellow to yellow-green fruiting bodies, which are usually no more than about ¾ to 2 inches (2 to 5 cm) tall, occur on soil or forest floor litter (PLATE 30). *Leotia viscosa* (green-headed jelly club), a closely related species, is very similar but has a dark olive-green head.

Earth Tongues

Some of the fungi known as earth tongues belong to the Helotiales, but others that were once placed here have been reassigned to other orders. All of these fungi produce fruiting bodies that are stalked and somewhat flattened, and this resemblance to a tongue accounts for their common name. *Microglossum rufum* (orange earth tongue) is one of the more frequently encountered species throughout North America. Its fruiting bodies are bright orange to yellow and 1¼ to 2 inches (3 to 5 cm) tall. They are found on decaying logs and leaf litter, often in association with mosses. Earth tongues that belong to orders other than the Helotiales include species in such genera as *Geoglossum*, *Spathularia*, and *Trichoglossum*.

Order Hypocreales

The order Hypocreales consists of a very large group of fungi, with at least several thousand species having been described. Some of these are known only from the asexual stage, so the type of fruiting body (a perithecium) characteristic of the order as a whole is not produced. Unlike many of the fungi with perithecia, members of the Hypocreales often produce pale or brightly colored (often yellow, orange, or red) perithecia that stand out in stark contrast to the usually darker substrates upon which they occur. Perithecia tend to be somewhat waxy or fleshy in texture and may occur singly or in dense clusters. When the latter is the case, the perithecia are often embedded in a common stroma.

Lobster Mushroom Fungus

Many of the taxa that have been assigned to the Hypocreales are saprotrophs, but some are parasites of plants and even of other fungi. The latter is the case for *Hypomyces lactifluorum* (lobster mushroom fungus), a parasite of the genera *Russula* and *Lactarius* in the Basidiomycota. When *H. lactifluorum* infects the fruiting body of one of these fungi, it produces a mycelium that completely covers the surface of the fruiting body. Embedded within the mycelium are thousands of bright orange-red perithecia, which cause the entire fruiting body to appear bright orange to orange-red in color (PLATE 31). Such infected fruiting bodies are referred to as lobster mushrooms because of their color. The presence of *H. lactifluorum* interferes with the normal development of the fruiting body, and the resulting somewhat funnel-shaped structure does not retain the features needed for identification. As will be reinforced several times later in this book, no fungus should ever be consumed unless it has been identified with absolute certainty. Nevertheless, lobster mushrooms are considered edible by some people and are often sold in the market.

Tree Pathogens

Species of *Nectria* are among the most frequently encountered members of the Hypocreales. Most are saprotrophs or weak parasites of woody plants, but a few are deadly pathogens that kill the host. These fungi often produce bright red, superficial perithecia that break through the bark of the infected host plant. *Nectria cinnabarina* (sometimes called coral-spot) is a common saprotroph on the twigs and branches of many different kinds of trees and other woody plants. Interestingly, both the perithecia and the conidia of the anamorph often occur simultaneously in infected areas of the host plant. *Nectria coccinea* causes beech bark disease on American beech (*Fagus grandifolia*). Infected trees are often killed by the fungus, and considerable loss of beech has occurred in some areas of eastern North America.

Order Laboulbeniales

The order Laboulbeniales is made up of what are clearly some of the most unusual ascomycetes, and a strong argument could be made that they constitute the most distinctive group of fungi. Indeed, when first described, these organisms were not always recognized as being fungi. Some species were first thought to be parasitic worms.

This order is the largest and most diverse group of fungi that occur as parasites of insects. All of the nearly 2000 species described to date are invariably attached to the exoskeleton of insects or (more rarely) millipedes or mites.

These fungi generally lack a mycelium and the entire vegetative body (thallus) is derived from the enlargement and subsequent cell divisions of a two-celled ascospore (PLATE 32). The growth of the thallus is determinate in that once a certain size is reached, growth stops. A peglike or rootlike haustorium is located at the very base of the thallus, and this penetrates the body of the host. There is little actual disruption of the tissues of the host, which seems to be relatively unaffected by the presence of the fungus.

Worldwide, the Laboulbeniales are currently known to parasitize 10 different orders of insects, but the vast majority of species are associated with beetles and flies. In general, they seem to occur with the highest frequency on insects from riparian (streamside) and aquatic habitats.

In the Laboulbeniales an anamorph is not produced, and reproduction is by asci and ascospores. These are produced in a perithecium that develops as an extension of the thallus. A median septum develops in each ascospore, dividing it into a two-celled structure. Upon being released, the ascospore becomes attached to a potential host and then germinates, eventually giving rise to a new thallus.

In general, species in the Laboulbeniales display a considerable degree of specificity in terms of their hosts, and some species appear to be restricted to just one place on the body of the host. Since any thalli that might be present are collected along with the host insect, surveys for these fungi often involve going through numerous specimens of insects collected by entomologists and available at various research institutions and universities. The thalli can be observed with the aid of a hand lens and appear as tiny, spinelike projections from the exterior of the host.

Members of the Laboulbeniales are certainly not uncommon, but they have been studied by relatively few mycologists. Roland Thaxter, a mycologist at Harvard University, devoted much of his career to these fungi and produced a monumental monograph published in five volumes that appeared from 1896 to 1931. Rarely has so much been written about what to even most mycologists is a rather obscure group.

Order Ophiostomatales

Some orders are more or less defined by one especially prominent member, and this is the case for the relatively small order Ophiostomatales, with no more than about 130 species. Members of the Ophiostomatales are characterized by perithecia with unusually long, slender necks. These are produced singly from the mycelium, which infects the wood and bark of particular types of host trees. Most species are associated with bark-boring beetles that disperse ascospores and conidia to new hosts. Although primarily saprotrophs, a few examples are deadly pathogens of trees.

Dutch Elm Disease

The best-known and most studied species in the order is *Ophiostoma ulmi*, one of the fungi that cause Dutch elm disease. In *O. ulmi*, the perithecia are black, and the long, cylindrical neck has a ring of specialized, hairlike hyphae that surround the opening at the top. When mature, the asci inside the perithecia disintegrate, and ascospores ooze out the top, producing a slimy droplet that is held in place by the hairlike hyphae, which form what is essentially a tiny basket. The anamorph of this species (placed in the genus *Graphium*) produces conidia on conidiogenous hyphae that occur in tightly packed bundles called synnemata (singular: synnema). These synnemata are relatively long, and a slimy droplet containing the conidia is formed at the apex.

This is a example of a situation in which the sexual spores (ascospores) and the asexual spores (conidia) occur on reproductive structures that have a similar functions, although each is derived in a very different fashion. In both instances, a slimy droplet containing spores is placed in an elevated position where it is more likely to come into contact with a beetle. Beetles are attracted to the fungus by the highly volatile compounds given off by the mycelium and then feed upon the ascospores and conidia.

The perithecia and synnemata typically occur in the tunnels created by bark beetles, and the slimy droplets readily adhere to the bodies of these beetles. Since the beetles tunnel beneath the bark of any potential new host tree, the spores are immediately placed in an advantageous place to initiate a new infection. Once established, the mycelium of the fungus spreads throughout the tree, where it interferes with the transport of water. An infection first becomes apparent when the leaves on an upper branch of the tree begin to wilt in dry weather and then rapidly turn brown and brittle. This progressively spreads to the rest of the tree, with further dieback of branches. Eventually, the entire tree dies. However, *Ophiostoma ulmi* is able to persist as a saprotroph in the bark of the dead trees.

Although *Ophiostoma ulmi* is directly responsible for the death of a tree, Dutch elm disease is usually referred to as a disease complex since the fungus involved is invariably associated with another organism (the bark-boring beetle). This same term is used for several other important tree diseases, including beech bark disease. For Dutch elm disease, the beetles play an essential role by first serving as a vector for spores and then by disrupting the tissue of the host tree so it can be easily infected by the fungus.

Ophiostoma ulmi is believed to have been accidentally introduced into both North America and Europe from Asia. The disease was first described in the late 1920s in the Netherlands, and its common name reflects this. Dutch elm disease had reached North America by the 1930s. It affects all species of elm, but American elm (*Ulmus americana*) is particularly susceptible.

Although Dutch elm disease had a considerable impact on populations of American elm throughout North America during the first three decades following its introduction, this was nothing compared to the almost complete devastation caused by a more aggressive form of the disease that was first discovered in the mid-1960s. This more aggressive disease is caused by what has come to be regarded as a separate species, *Ophiostoma novo-ulmi*. American elm, which used to be common in urban settings, has been virtually eliminated in many parts of its former range.

Blue-Stain of Wood

Ophiostoma piceae, a related species, is responsible for blue-stain of wood. This fungus, just like *O. ulmi* and *O. novo-ulmi*, is invariably associated with wood-boring beetles. It is most commonly encountered in conifers, including various species of pine. Unlike *O. ulmi* and *O. novo-ulmi*, *O. piceae* is much more likely to be found in a log on the ground or a dead-but-still-standing tree than in a living tree. The presence of the fungus causes a bluish or grayish discoloration of the infected wood but does not affect its strength. In a cross section of a log, the discoloration often appears initially as a pie-shaped wedge, but over time it may extend to all of the outermost wood and appear as spots, streaks, or patches elsewhere. Although the fungus does no real damage, the discoloration itself lowers the value of the wood, which is a serious financial issue to the timber industry.

Order Pleosporales

Like the Capnodiales, the order Pleosporales is assigned to the assemblage commonly referred to as the loculoascomycetes. The distinguishing feature of members of this assemblage is the production of a type of fruiting body known as a pseudothecium (plural: pseudothecia). Within the pseudothecium, asci develop within cavities called locules. The latter provide the basis of the name used for these fungi. In addition, asci have two wall layers, a condition referred to as bitunicate (the word essentially translates as "two coverings").

Fruiting bodies produced by loculoascomycetes are variable in form and may be unilocular (contain only a single cavity or locule) or multilocular (contain many locules). When the former condition exists, the fruiting body closely resembles a perithecium. The order Pleosporales is extremely large (at least 6300 species) and morphologically diverse.

Black Knot of Cherry

Apiosporina morbosa (black knot of cherry) is one of the more commonly encountered examples of the order Pleosporales. This fungus infects species of cherry and other closely related members of the Rosaceae, where it results in the formation of a conspicuous black, knotlike structure (hence the common name) on the branches of the host tree (PLATE 33). There appears to be considerable variation in the level of susceptibility of the host. Sometimes, the fungus seems to have little effect on the tree, but in other instances infected twigs and branches die. Severe infections can result in the death of the entire tree. In winter, after trees have lost their leaves, the identification of most species of trees is often problematic. In eastern North America, wild cherry (*Prunus serotina*) is easy to identify, even at a considerable distance, because of the presence of the unsightly black knots.

Apple Scab

Venturia inaequalis, the fungus that causes the disease known as apple scab, is the member of the Pleosporales most likely to be familiar to the average person. *Venturia inaequalis* occurs throughout the world wherever apple (*Pyrus malus*) and other closely related members of the Rosaceae such as hawthorn (*Crataegus*) are found.

An infection of an apple tree begins in the spring when an airborne ascospore of *Venturia inaequalis* adheres to the moist surface of a young leaf. Germination of the ascospore requires the presence of moisture, so periods of infection coincide with periods of rainy weather. The germinated ascospore produces a mycelium that forms a thin layer below the cuticle of the leaf. Within a few days, the mycelium gives rise to numerous short conidiophores that break through the cuticle. The conidiophores produce conidia that can be spread by wind or rain to other leaves or the developing apple fruits.

Late in the growing season, the mycelium of the fungus penetrates deeper into the leaf and begins to produce pseudothecia. The development of these structures requires several months and is not completed until well after the leaves have fallen to the ground. In the spring, the asci reach maturity and the ascospores are released. These can potentially infect new hosts.

The infected areas (referred to as scabs) on an apple caused by *Venturia inaequalis* are rather superficial, and the apple itself is still completely edible. However, the market value is considerably reduced, since many consumers would not purchase such an "unattractive" apple. The impact of the fungus on the rest of the tree is more significant. If the disease is not controlled, the overall yield is reduced. Some apple growers in the United States have been reported to lose up to 70 percent of their apple crop to this disease in some years.

Order Sordariales

The Sordariales, the last of the orders of ascomycetes considered in this chapter, is a group of at least 500 species of mostly coprophilous ("dung loving") fungi. Some examples occur on wood or soil. For example, the latter is the case for members of *Neurospora*, the best-known and most studied genus in the order.

Fungi on Burnt Plant Matter

Most species of *Neurospora* occur on soil and are among the first fungi to be found on burned-over areas of soil and charred plant material following fire. In areas of the tropics where forests are cut and then deliberately burned to clear land for growing crops or grazing livestock, the burned areas are often covered by mycelial mats that appear pink or orange from the enormous numbers of conidia that are produced. Apparently, dormant ascospores in the soil are stimulated to germinate by heat and produce a rapidly growing mycelium. Because it is easy to grow and maintain in laboratory culture, certain species of *Neurospora* has been used extensively as experimental organisms for studies of fungal genetics and biochemistry. In fact, our understanding of the relationship that exists between proteins and genes in living systems was first developed from studies that used *N. crassa* as an experimental organism.

Dung-loving Ascomycete Fungi

Members of the genus *Sordaria* are common on the dung of herbivores. *Sordaria fimicola* is especially common on horse dung, where its solitary perithecia are scattered across the surface of the dung. Ascospores are forcibly discharged from the perithecia. In *Sordaria*, the neck of the perithecium is positively phototropic and thus directed towards the light. As the asci in the perithecium mature, they swell and fill the upper portion of the perithecial cavity. Eventually, one ascus extends upwards through the neck of the perithecium far enough for its tip to protrude from the opening at the top. Soon after this happens, the ascus explosively discharges all of the ascospores and then collapses and disintegrates. One after another, the other asci in the perithecium follow the same sequence of events.

The small dark perithecia of species of *Podospora* are often exceedingly common on the dung of large herbivores (PLATE 34). In some members of the genus, the outer surface of the perithecium is covered with scattered bristlelike hyphae. One unusual feature of species of *Podospora* is that each ascospore has a mucilaginous appendage.

Fabric- and Paper-loving Fungi

In members of the genus *Chaetomium*, ascospores are liberated from the perithecium in quite a different manner. These fungi, which are often associated with cellulose-rich substrates such as fabrics and paper, have asci that are evanescent, leaving behind a mucilaginous mass containing the ascospores in the cavity of perithecium. When the ascospores are liberated from the latter, they literally ooze out, often forming contorted masses that become trapped in the numerous long, hairlike hyphal appendages that decorate the outer surface of the perithecium.

5
Morels, Truffles, Cup Fungi, and Flask Fungi

True Morels 85	**Cup Fungi** 92
False Morels 87	Family Pezizaceae 93
True Truffles 88	Family Sarcosomataceae 93
Morphology of Truffles 88	Family Pyronemataceae 93
Hunting Truffles 89	Family Ascobolaceae 94
Truffles in History 90	**Flask Fungi** 95
Two Noteworthy Truffles 91	Family Xylariaceae 96
Decline of Truffles in Nature 91	Family Diatrypaceae 97
False Truffles 92	**The Asa Gray Disjunction for Fungi** 99

Two orders (the Pezizales and the Xylariales) in the Ascomycota contain numerous taxa that characteristically produce fruiting bodies large enough to be easily noticed in the field. In the Xylariales, a perithecium is produced, whereas in the Pezizales, an apothecium is produced. The order Pezizales includes a number of different families. The more important of these are the Ascobolaceae, Helvellaceae, Morchellaceae, Pezizaceae, Pyronemataceae, Sarcosomataceae, and Tuberaceae. Perhaps the best known of these to the average person is the Morchellaccac, which consists of morels.

True Morels

In the true morels, the fruiting body consists of an upper more or less conical portion (cap) held aloft on a broad, hollow stalk. The cap

85

appears honeycomb-like in that it is composed of a network of ridges with pits between the ridges (PLATE 35). Since the hymenium is confined to the pits, it seems likely the entire fruiting body of a morel is a composite structure that actually consists of multiple fruiting bodies, with each pit representing an apothecium.

Morels are highly prized as edible fungi and are hunted by thousands of people every year simply for their taste and the joy of the hunt. For many people, morels are the only wild fungi collected for the table. Morels have been given many local names, some of the more colorful of which include dryland fish (due to their similarity in taste to fish), sponge mushrooms (because of their general appearance), molly moochers (a name used mostly in parts of the Southern Appalachians), or merkels (essentially a corruption of the word "miracles" and based on a story of how a mountain family was saved from starvation by eating morels).

Although edible, morels do contain small amounts of toxins usually rendered harmless thorough cooking. As such, morels should never be eaten raw. There have been reports that even cooked morels can sometimes cause a mild case of poisoning when consumed along with alcohol.

The most commonly encountered morels are members of the genus *Morchella* (from *morchel*, an old German word for mushroom), which includes at least a dozen different species in North America and an unknown number worldwide. Some of the common forms in North America are *M. esculenta* (yellow or common morel), *M. deliciosa* (white morel), *M. elata* (black morel), and *M. semilibera* (half-free morel). *Morchella semilibera* is the only true morel in which the lower margin of the cap is not attached directly to the stalk.

People who collect morels tend to refer to the various species they encounter by their color (for example, white, tan, gray, yellow, or black). Black morels are the most distinctive and are often the first morels to appear in the spring (PLATE 36).

Because morels are so highly prized, particular spots that are especially productive for these fungi tend to be kept as jealously guarded secrets by collectors. Just what causes one spot to be better than another for morels is not known, although collectors have long noted the apparent association of morels with certain types of trees, especially tulip-poplar (*Liriodendron tulipifera*) and American ash (*Fraxinus americana*). Apple orchards, particularly those with old trees, are reputed to be productive for morels, as are recently burned areas. In fact, morels often occur abundantly in the two or three years immediately following a forest fire, and commercial collectors in places like Alaska are known to follow forest fires to take advantage of the bountiful collecting that is often possible.

Morels have not yet been successfully cultivated on a large scale, and the commercial morel industry is based largely on collecting specimens in nature.

False Morels

When collecting morels, care must be taken to distinguish them from false morels, some of which are poisonous. The false morels belong to the family Helvellaceae. Although people do sometimes confuse the fruiting body of a false morel with that of a true morel, the two are actually quite different. Like morels, the fruiting body of a false morel is divided into a stalk and a head. The head is rather variable and can be either wrinkled or smooth, with an overall form ranging from conical to ovoid or even cup-shaped. The cap never has the system of ridges and pits so characteristic of the true morels.

Members of two genera (*Gyromitra* and *Helvella*) are the most frequently encountered false morels. Species of *Helvella*, which are sometime referred to a saddle fungi, have a fruiting body in which the cap is a curved, saddle-shaped structure that looks as if it had been folded over the stalk.

Gyromitra brunnea (brown false morel) is one of the larger (with fruiting bodies sometimes more than 6 inches [15 cm] tall) and more common representatives of the genus *Gyromitra* (PLATE 37). In this species the cap is irregularly lobed or folded and then somewhat saddle-shaped. *Gyromitra brunnea* is one of the first fleshy fungi to appear in the spring and often appears about the same time as the first true morels.

Some species of *Gyromitra* contain the toxin gyromitrin, which is converted to the even more deadly monomethylhydrazine. If consumed, these toxins can break down blood cells and cause acute liver damage. Gyromitrin also has been shown to be highly carcinogenic in small animals. Although this fact is generally known, some people continue to collect and eat these fungi. Cooking apparently drives off some of the toxic substances, which improves the odds of not being affected. However, *G. brunnea* and other related forms such as *G. esculenta, G. gigas,* and *G. infula* are still far too dangerous to be recommended for human consumption.

True Truffles

Although morels are highly prized as edible fungi in certain circles, for some people (especially in Europe) morels do not compare with the "gastronomic treasures" found in another group of fungi—the truffles. The members of the order (Tuberales) to which truffles belong are distinctive because they produce fruiting bodies that are subterranean (that is, occur underground). Because of this, they are rarely observed in nature.

Subterranean (or hypogeous) fruiting bodies are not limited to the Tuberales and (as will be mentioned elsewhere in this book) have evolved in several groups of fungi, including some members of the Basidiomycota. It has been suggested that they represent an adaptation for surviving drought, forest fires, or extreme cold, but some examples of hypogeous fungi are found in areas of the world where such conditions rarely occur.

Morphology of Truffles

Because there are other kinds of hypogeous fungi, it is important to note that only the "true truffles" are members of the Tuberales. The most important genus is *Tuber*, for which about 100 species have been described.

Truffles are mycorrhizal fungi associated with a number of different kinds of trees, including beech, hazelnut, oak, and pine. The fruiting body tends to be roughly globose, although its shape is influenced by the surrounding soil and rocks in which the structure develops. Depending upon the species involved, fruiting bodies vary in size from less than ½ inch (1.3 cm) to more than 2 inches (5 cm) in diameter, and exceptionally large examples can weigh as much as 35 ounces (1000 grams).

When mature, the fruiting body is rather dense and firm to even hard to the touch. This is rather unlike the fruiting bodies of most other fungi but perfectly understandable under the circumstances, since a soft and fragile fruiting body would be poorly equipped to develop in what is often rather hard soil. The interior of the fruiting body is differentiated into an outer sterile layer (called a peridium) that completely surrounds an inner fertile portion (referred to as the gleba). Within the gleba, asci are produced within a highly contorted hymenium that is quite distinctive. The individual asci are thin-walled, rounded and usually contain only one to three ascospores. The latter tend to have highly ornamented spore walls but no obvious means of being liberated from the ascus. Indeed, truffles have evolved a system of spore dispersal that is totally dependent upon the activities of certain insects and small mammals.

In brief, when the ascospores are mature, the fruiting body begins emitting what are best described as fascinating (albeit not always pleasant) odors. These odors, which result from various highly volatile organic compounds produced by the truffle, attract insects and small mammals. The insects involved are usually small flies (members of the genus *Suillia* and known as truffle flies) that locate and then lay their eggs in the truffle. After the eggs hatch, the fly larvae feed upon the truffle. From the standpoint of the truffle, this is much less desirable than being located by a small mammal such as a badger, mouse, shrew, or squirrel. These excavate and consume the truffle, although usually scattering some ascospores in the process. More importantly, the mammal ingests numerous ascospores, which pass unharmed through its digestive tract and are deposited elsewhere. In many species of truffles, germination of the ascospores is thought to be enhanced by passage through the digestive tract of the mammal vector.

Hunting Truffles

In some parts of Europe, dogs and pigs have been trained to locate truffles. The use of pigs (usually female pigs or sows) might seem unexpected, but recent biochemical studies have revealed that one of the substances emitted by truffles is a steroid hormone identical to the main sex hormone produced by male pigs (or boars). Presumably, this hormone allows the sow to respond to truffles by instinct alone and thus reduces the amount of training required.

Some truffle hunters find truffles by looking for truffle flies, a technique known in France as *avec la mouche* (literally "with the fly"). This method usually involves walking through an areas where truffles are thought to be present and using a long stick to disturb the ground litter and low-growing plants. Numerous truffle flies tend to occur in places where truffles are present beneath the surface of the ground, and these can be observed to fly away when disturbed. Amazingly, some humans have a keen enough sense of smell to detect truffles just by sniffing the ground.

Truffles in History

Truffles have been known to humans for at least several millennia and probably much longer. References to what were unquestionably truffles first appear in the writings of the Greek philosopher Theophrastus (370 to 286 BC). They were also mentioned in the works of such early writers as the Roman military commander and natural philosopher Gaius Plinius Secundus, who was better known as Pliny the Elder (23 to 79 AD), and the Greek physician and pharmacologist Pedanius Dioscorides (40 to 90 AD).

Because truffles occurred in the ground but had no obvious stem or root, their origins were a mystery. In fact, a number of what today would be regarded as rather fanciful explanations (for example, that truffles were the products of thunder and lightning) were advanced.

There is little doubt that the ancient Greeks and Romans shared the same high culinary regard for truffles found among many modern humans, but the truffles consumed were not necessarily from the same taxa. When writing about truffles, Pliny the Elder noted that "those of Africa are most esteemed," and it seems very likely that he was referring to the so-called desert truffle (genus *Terfezia*) and not the species of *Tuber* described below.

Terfezia is not closely related to *Tuber*, but does consist of a number of edible species. These occur in arid regions of southern Europe and the Middle East, where they form ectomycorrhizal associations with certain shrubs, including members of the genus *Helianthemum*. It seems likely that desert truffles have been collected and consumed by humans for thousands of years, and these fungi can still be found in local markets throughout the regions where they occur.

Two Noteworthy Truffles

Tuber melanosporum (Périgord black truffle), which derives its common name from the Périgord region in France, is generally regarded as the most exquisite of all truffles, although not necessarily the most valuable. Fruiting bodies of this species, which are found associated with oak, range in size from little more than the size of a fingernail to larger than a tennis ball, sometimes weighing more than 7 ounces (200 grams). The surface, which is covered in small diamond-shaped facets, is a dark ruddy brown when young but becomes black when the fruiting body is fully mature (PLATE 38). Because this truffle is so very expensive, gourmet cooks use only small pieces to enhance the flavor of their dishes instead of serving it up as a separate part of a meal.

Tuber magnatum (Italian white truffle) typically commands the highest price of any truffle, sometimes more than U.S. $1000 per pound (ca. €1670 per kg in 2009). This species is associated with a number of different trees, including oak, hazelnut, poplar, and beech. The fruiting bodies have a smooth to somewhat suedelike surface, and the color of the peridium ranges from pale yellowish brown to greenish gray, with black, rusty, or brown spots. When sectioned, the gleba is pale cream or brown with white marbling. Fruiting bodies can exceed 4 inches (10 cm) in diameter and weigh more than 17 ounces (500 grams), although they are usually much smaller.

Decline of Truffles in Nature

Throughout those regions of Europe where at least some records are available for the numbers of truffles collected each year, a disturbing trend is apparent. Far fewer truffles are being collected today than a century or so ago. There are probably a number of factors that have contributed to this decline. These include both changes that have taken place in the landscape (for example, marked increases in the extent of urban areas) and in society (for example, loss of the expertise needed to collect and cultivate truffles as more people have moved to urban areas).

The lower numbers of truffles coupled with an ever-expanding market have prompted efforts to produce these fungi in other parts of the world, and several of these have achieved some degree of success. The first Périgord black truffles to be produced in the Southern Hemisphere were harvested in New Zealand in 1993, and in 1999, the first truffles were harvested in Tasmania. Readers interested in learning more about these efforts or any other aspect of truffles should take a look at the book *Taming the Truffle* by Ian Hall, Gordon Brown, and Alessandra Zambonelli, published by Timber Press in 2007.

False Truffles

Species of *Elaphomyces* are the most common hypogeous fungi in temperate regions of the world (PLATE 39). These fungi belong to the order Eurotiales (described in the previous chapter) and not the Pezizales. As such, they would be referred to as false truffles.

The fruiting bodies of *Elaphomyces* are distinctly different from those of true truffles. When mature, they consist of a two-layered peridium and a central yellow-brown to black powdery mass of ascospores. These fruiting bodies range from ³/₈ to 1¹/₄ inches (1 to 3 cm) in diameter and form an important item in the diet of some small mammals such as the northern flying squirrel. The whitetail deer is known to dig up and consume *Elaphomyces* species that occur in eastern North America, which are often referred to as deer truffles. These fungi can be abundant at times in red spruce forests in the Appalachian Mountains of West Virginia.

Since *Elaphomyces* and other hypogeous fungi are not usually visible in nature, collectors often search for their fruiting bodies with a small hand rake (usually called a "truffle rake"). The rake is used to pull away the uppermost layer of leaf litter, humus, and soil so that any fruiting bodies present will be revealed. When the area to be surveyed for hypogeous fungi is characterized by a plentiful supply of tree roots, this requires considerable effort. However, the discovery of fruiting bodies, several of which often occur together in one small area, makes it worthwhile.

Cup Fungi

The majority of what most people would refer to as "cup fungi" belong to the order Pezizales. Many of the more than 1100 species of fungi in the order produce an fruiting body that is usually recognizable as an apothecium. The apothecium often takes the form of saucer, cup, or bowl, with the hymenium occurring as a layer over the upper surface. The usual situation is for an apothecium to lack a stalk, but the latter is found in some species. Apothecia can range from less than 1/32 inch (1 mm) to more than 2 inches (5 cm) in total extent. Most occur as saprotrophs on soil, dung, or decaying wood.

Family Pezizaceae

Some of the more typical members of the family Pezizaceae are found in the genus *Peziza*, which consists of about 100 species. The usually pale brown and sometimes irregularly cup-shaped fruiting bodies of these fungi typically occur on decaying wood or humus-rich soil. In most species of *Peziza*, the fruiting body is relatively large, sometimes reaching 1 to 2 inches (2.5 to 5 cm) in diameter. *Peziza phyllogena* (common brown cup) can sometimes reach 4 inches (10 cm) in diameter.

Family Sarcosomataceae

The family Sarcosomataceae includes a number of taxa in which the fruiting body is often stalked and not as fleshy as most members of the Pezizaceae. Some examples are exceedingly colorful. This is the case for *Sarcoscypha occidentalis* (stalked scarlet cup), in which the fertile upper surface of the shallow cup-shaped fruiting body is a bright scarlet red (PLATE 40). The cup is borne on a white or pinkish white stalk that is about as long as the cup is wide. Although relatively small (usually less than 1/2 inch [1.3 cm] in diameter), *S. occidentalis* is easily seen on the forest floor, where it occurs in small clusters on twigs and small dead branches. After a period of rainy weather, I have observed this fungus to be so abundant in some upland broadleaf forests in the Appalachians that almost every twig seemed to be decorated with the tiny scarlet cups.

Family Pyronemataceae

Another frequently encountered cup fungi is *Scutellinia scutellata* (eyelash cup), the fruiting bodies of which often occur in some abundance on moist decorticated wood. In most forests throughout North America, it is often possible to find this fungus (a member of the family

Pyronemataceae) on just about any well-decayed log in which the wood remains moist. The fruiting bodies of *Scutellinia scutellata* can be distinguished from those of other morphologically similar fungi because the edges of the small (usually no more than ½ inch [1.3 cm] in diameter), shallow, red to bright orange-red cups have radiating long dark hairs that resemble eyelashes (hence the common name).

Scutellinia scutellata occasionally occurs on humus-rich soil, and this is the usual substrate for some of the other species in the genus (PLATE 41). When I spent late January to early May of 1995 at an Australian Antarctic Division Research Station on subantarctic Macquarie Island, the bright red fruiting bodies of *Scutellinia* were consistently abundant in most of the habitats surveyed for fungi.

While the fruiting bodies of *Scutellinia* have the shape of shallow cups, this is not the case throughout the entire family. In fact, there are other types of fruiting bodies found in the Pyronemataceae that probably reflect two of the steps involved in the evolution of the highly specialized hypogeous fruiting body characteristic of the true truffles. For example, species in the genus *Genea* produce closed but hollow fruiting bodies, while those in the genus *Geopora* produce fruiting bodies in which the hollow spaces are much reduced. The fruiting body of a true truffle is solid and there are no hollow spaces.

The fruiting bodies of both *Genea* and *Geopora* are hypogeous, and it is easy to see how the evolutionary process could begin with a fruiting body having an upturned margin (*Scutellinia*), proceed to one in which the margins have folded over to enclose the hymenium (*Genea*), continue to examples in which the open spaces within the fruiting body are largely eliminated (*Genea*), and finally end up with the solid structure of a true truffle (*Tuber*). The fruiting bodies of species of *Genea* are not uncommon, but they are too small (usually less than ⅜ inch [1 cm] in diameter) to be of interest for human consumption. When cut in half, their convoluted nature is somewhat suggestive of small human ears.

Family Ascobolaceae

The best place to look for members of the family Ascobolaceae is the dung of herbivores, because this is a largely coprophilous family. Species of *Ascobolus* are particularly common on dung, usually appearing when the latter is still relatively fresh. These species produce small, pale yellow or somewhat translucent apothecia that often occur in large numbers over the surface of the dung. When mature, each apothecium appears to be studded with small purple dots. Each dot is the projecting tip of an ascus, with the color derived from the conspicuous red to purple ascospores.

When they first develop, the asci are colorless and occur within the hymenium that covers the upper surface of the apothecium. However, as the asci mature they elongate so that their tips project well above the hymenium. The tips are positively phototrophic and as elongation occurs, they become oriented in the direction of the incoming light (PLATE 42). The ascospores are forcibly discharged and can travel far enough to reach plants growing beyond the limits of the dung. The surface of an ascospore is sticky, which causes the ascospore to adhere to any plant surface upon which it happens to land. If consumed by a herbivore, the ascospores survive passage through the digestive tract and are deposited as part of the dung.

Flask Fungi

The fruiting bodies produced by members of the order Xylariales are not nearly as well known as those produced by members of the Pezizales, although they tend to be much more common in nature. Most examples are not very conspicuous and are easily overlooked in the situations in which they occur. The majority of the more than 800 described species in the order are saprotrophs and typically occur on the bark and wood of still-standing dead trees, fallen logs, and stumps. Others are plant pathogens, and a few of these are economically important.

The fruiting bodies produced are most commonly perithecia, and this would (as described in Chapter 4) place the Xylariales in the assemblage of fungi referred to as pyrenomycetes. In fact, some mycologists restrict the use of the term "pyrenomycetes" to just this one order. Individual perithecia in members of the Xylariales are relatively small and most are dark in color. In some taxa, they occur singly, but the usual situation is for perithecia to be grouped together in a common stroma (plural: stromata). The latter is often a dark and carbonaceous structure within which the perithecia are more or less embedded.

Because of their dark color, these stromata do not stand out against the wood or bark substrates upon or within which they occur. However, anyone who takes the time to examine such substrates is likely to discover that this group of fungi is well represented.

Since most stromata are tough or hard enough to be fairly resistant to decay, they may persist for several months after the ascospores have been discharged. This contrasts with the fruiting bodies produced by members of the Pezizales, which usually do not last for very long in nature after they have become mature.

Members of two families in the Xylariales are commonly encountered in nature. These two are the Diatrypaceae and the Xylariaceae, with fungi belonging to the latter generally being more conspicuous and thus better known.

Family Xylariaceae

Taxa assigned to the Xylariaceae are characterized by a type of ascus with a distinct pluglike apical ring that stains blue in the presence of iodine. The ring appears to have some effect upon the way in which ascospores are discharged from the ascus, but whether or not this is an important factor for the fungus is still not completely clear. Three of the more distinctive and widespread genera within the Xylariaceae are *Xylaria, Daldinia,* and *Hypoxylon*.

Xylaria is a genus of at least 100 species, most of which occur as saprotrophs on decaying wood. One of the more common species is *X. polymorpha*, commonly known as dead man's fingers because the fruiting bodies that project from the ground often resemble fingers (PLATE 43). Each fruiting body is a composite structure consisting of hundreds of individual perithecia embedded in an elongated stroma. The surface of the stroma initially appears white and powdery because of the presence of conidiophores and conidia, but it later turns black, looking much like a piece of charred wood. Breaking one of the "fingers" reveals that the stroma has a pure white interior. In *X. polymorpha*, it is usually possible to discern the outer boundary of the mycelium in a log or stump due to the presence of a conspicuous black line of demarcation in the wood.

The more than 120 species of *Hypoxylon* commonly occur on decaying bark and wood, where their often hemispherical or somewhat flattened stromata sometimes persist for several years. Although usually regarded as saprotrophs, some of these fungi infect still living trees, which would mean that they are at least weak parasites of the infected trees. The genus *Hypoxylon* was once considered to encompass a number of taxa that have now been placed in other genera on the basis of microscopic features of the stroma and/or ascospores. For example, several species formerly considered to belong to *Hypoxylon* have been reassigned to such genera as *Biscogniauxia* and *Camillea*. Such a thing is not unusual and has happened for numerous groups of fungi. In theory, this results in a more homogenous taxonomic assemblage for a particular taxon at whatever taxonomic level is involved.

Some species of *Hypoxylon* display a preference for wood from a particular host tree. This is the case for *H. fragiforme* (red cushion hypoxylon), which is restricted largely to dead branches of beech (*Fagus*). Fruiting bodies (actually stromata) of this fungus are $1/16$ to $5/8$ inch (1.5 to 15 mm) in diameter, occur in clusters (PLATE 44), and are usually easy to spot because of their color. Each fruiting body is more or less globose and salmon-pink at first, but it eventually becomes brick red in color, hence the common name.

The fruiting bodies of species of *Daldinia* are easily recognized because they have concentric zones of alternating light and dark bands in the interior. In fact, the species name for the very common *D. concentrica* (carbon balls) is based on this feature, which is strikingly apparent when a fruiting body is cut in half. The fruiting bodies of *D. concentrica*, which are hard and black, occur on dead branches and stumps of broadleaf trees, either individually or in small clusters (PLATE 45). Most fruiting bodies are $3/4$ to $1 5/8$ inches (2 to 4 cm) in diameter. They are apparently very effective in storing moisture, because a mature fruiting body will continue to actively discharge ascospores even when subjected to exceedingly dry conditions. One unusual feature is that the ascospores are discharged almost exclusively at night, and a single fruiting body can produce ten million spores each night for several weeks (Boddy et al. 1985).

Family Diatrypaceae

Members of the Diatrypaceae are common but often overlooked. Most are saprotrophs and inhabit bark or wood, but a few are pathogenic. For some species in this family, the ascus has an apical ring that stains blue in the presence of iodine, a feature shared with the Xylariaceae. However, the ascospores produced are characteristically sausage-shaped, a feature that serves to distinguish the Diatrypaceae from the Xylariaceae. Among the more common and widespread examples of fungi that belong to the Diatrypaceae are species in such genera as *Diatrype* and *Diatrypella*.

Within the Diatrypaceae, taxa are distinguished on the basis of both readily apparent morphological characteristics and less obvious microscopic features. For example, the genera *Diatrype* and *Diatrypella* differ primarily in the number of ascospores produced per ascus, with eight for *Diatrype* and many (a condition referred to as polysporous) for *Diatrypella*. Moreover, most species of *Diatrype* produce stromata that extend over large areas of the substrate upon which they occur, while in *Diatrypella* the stromata are much smaller and usually wartlike or discoid.

However, discoid stromata are sometimes found in certain species of *Diatrype* such as *D. virescens*, which has distinctive yellow-green stromata and is restricted to beech in North America (PLATE 46). Another discoid species known from beech is *D. disciformis*, which is common in Europe.

Interestingly, the species of *Diatrype* characterized by extensive stromata are often restricted to host plants in a single genus. Prominent example are *D. decorticata* on beech, *D. undulata* on birch, and *D. stigmaoides* on oak. Species of *Diatrypella* also exhibit a tendency to be associated with certain host plants, and many are restricted either to a single genus (for example, *D. quercina* and *D. pulvinata* on *Quercus* [oak]) or to a family, such as *D. placenta* on members of the Betulaceae [birch family]).

Other genera assigned to the Diatrypaceae differ from *Diatrype* and *Diatrypella*, both of which have stromata that break through the bark as compact mycelial masses, by producing stromata within the tissues of the host. For example, the stromata of *Cryptosphaeria* develop just beneath the bark, while those of *Eutypa* usually occur within the wood but also can sometimes penetrate the bark. As such, these genera essentially inhabit different "life zones" within the host plant.

Pathogenic species in the Diatrypaceae are very rare, with the most noteworthy exceptions being some species of *Eutypa* that are regarded as the causal agents of the canker diseases of apricot and olive trees.

The Asa Gray Disjunction for Fungi

The nineteenth-century plant taxonomist Asa Gray (1810–1888) noted that a number of morphologically similar plants occurred only in East Asia and eastern North America. Some of the same species and 65 genera are found in the two regions. Prominent examples included *Panax* (ginseng), *Cornus* (dogwood), and *Podophyllum* (mayapple). This pattern has since become known as the "Asa Gray disjunction" and has been the subject of a number of studies (Wen 1999).

Since many fungi, including the members of the Diatrypaceae mentioned above, are usually associated with specific genera or families of host plants, it would seem likely that some fungi conform to this same pattern. Few studies of a possible "Asa Gary disjunction" for fungi have ever been carried out. Too often, the assumption has been made that most species of fungi have very large ranges and do not display clearly defined biogeographical patterns.

Larissa Vasilyeva, a mycologist at the Institute of Biology and Soil Science of the Far East Branch of the Russian Academy of Sciences, has collected fungi (including members of the Diatrypaceae) in both East Asia and the eastern United States. She has concluded that the distribution patterns of many of the fungi reported from these two regions should be investigated more closely to clarify the taxonomic concepts used for the species involved. In some cases, morphologically similar fungi present in both regions and considered to represent the same species may instead be two distinct species. In other cases, a single species not recognized as such may occur in both regions.

One of the best-known examples of the "Asa Gray disjunction" for fungi involves two species in the genus *Ciboria*. *Ciboria carunculoides* is found in eastern North America and *C. shiraiana* in eastern Asia. Both are parasites of mulberry and cause a condition called the "popcorn disease" in the fruits.

Another example is found in the recently described genus *Cryptovalsaria*, whose members are pathogens of alder. Although *Cryptovalsaria* is assigned to the Diatrypaceae, the ascospores it produces are two-celled, a condition not present elsewhere in the family. *Cryptovalsaria rossica* is found on alder in East Asia, while *C. americana* occurs on alder in eastern North America.

6
Mushrooms and Other Larger Fungi

Reproduction in Phylum Basidiomycota 101
Classification of Subphylum Agaricomycotina 102
Agaricales 104
Morphology of Agarics 104
Identification of Agarics 105
Agarics with White Spores 107
Family Amanitaceae 107
Family Lepiotaceae 108
Family Russulaceae 109
Family Hygrophoraceae 109
Family Tricholomataceae 110
Family Marasmiaceae 111
Family Mycenaceae 111
Agarics with Salmon to Pink Spores 111
Agarics with Black to Smoky Gray Spores 112
Agarics with Purple-Brown to Chocolate-Brown Spores 112
Agarics with Clay-colored Spores 113

Agarics with Bright Yellow-Brown to Clay-Brown Spores 113
Boletes 114
Aphyllophorales 115
Polypores 115
Corticioid Fungi 117
Chanterelles 118
Tooth Fungi 118
Coral Fungi 119
Gasteromycetes 120
Puffballs 120
Earthstars 121
Bird's Nest Fungi 122
Stinkhorns 123
Heterobasidiomycetes 125
Genus *Tremella* 125
Genus *Auricularia* 126
Genus *Pseudohydnum* 126
More About Basidiomycetes 127

The large and conspicuous fungi that one encounters in forests and fields are mostly macroscopic members of the phylum Basidiomycota. The defining characteristic of members of this phylum is the basidium (plural: basidia), a special type of hypha on which basidiospores are produced. This structure defines the Basidiomycota in the same way that the ascus defines the Ascomycota.

Up to a point, the development of a basidium is not unlike that of an ascus. The young basidium, like the young ascus, is dikaryotic, with two haploid nuclei present. The two nuclei fuse, resulting in a diploid nucleus. Meiosis then occurs to produce four haploid nuclei. While this is happening, four small extensions, called sterigmata (singular: sterigma), develop from the upper part of the basidium. The tips of these sterigmata expand, becoming what are known as basidiospore initials. Eventually, one of the four haploid nuclei moves into each of the basidiospore initials and the latter mature into basidiospores (PLATE 47).

Unlike the usual situation in the Ascomycota, in which eight ascospores are produced per ascus, only four spores are produced per basidium in most members of the Basidiomycota (or basidiomycetes, as the term is used in an informal sense). More importantly, the basidiospores are located outside the basidium, while ascospores are located inside an ascus.

In many instances, the mature basidium is a relatively simple, club-shaped structure, but this is not always the case. Moreover, not all basidia produce four basidiospores, each with a single nucleus. In *Agaricus bisporus*, the commercial mushroom mentioned in Chapter 1, only two basidiospores are formed, and each of these has two nuclei. In fact, the specific epithet (*"bisporus"*) of this fungus reflects the fact that only two basidiospores are produced. Some basidiomycetes produce as many as eight basidiospores per basidium, while others produce the more typical four, but each of these has two nuclei (Webster and Weber 2007).

Reproduction in Phylum Basidiomycota

As noted above, the hypha that gives rise to the basidium is dikaryotic, which is the same condition that exists for the developing ascus at a comparable stage, although in the Ascomycota, the dikaryotic condition is limited to the ascogenous hyphae. The situation in the basidiomycetes is quite different. When a basidiospore germinates, the mycelium that eventually develops from it usually has only one type of nucleus present in each cell-like compartment of the individual hyphae. As such, it is monokaryotic, not unlike the mycelium that develops when an ascospore germinates, as described in Chapter 4. In the basidiomycetes, this monokaryotic mycelium is known as the primary mycelium. It may persist for a period of time, but ultimately it encounters another compatible primary mycelium of the same species of fungus. The two primary mycelia fuse and give rise to a secondary mycelium, which is dikaryotic because of the presence of two different types of nuclei (one each from the two primary mycelia) in each cell-like compartment.

In the comparatively rare instances in which a binucleate basidiospore is involved, a secondary mycelium is produced directly from the basidiospore when the latter germinates. This dikaryotic secondary mycelium is the predominant stage in the life cycle of a typical basidiomycete and can continue growing almost indefinitely if environmental conditions remain favorable. This is in striking contrast to the short-term nature of the condition in the Ascomycota.

The secondary mycelium differs from the primary mycelia not only in being dikaryotic but also in structure. The most obvious difference is the presence, at least in many basidiomycetes, of clamp connections. These structures are formed when, during the growth in length by a particular hypha, the original cell-like compartment at its tip divides to produce two new compartments. The two nuclei also divide, and the clamp connection is simply a device to ensure that each of these new compartments ends up with one nucleus of each of the two types.

Ultimately, a portion of the secondary mycelium differentiates into a tertiary mycelium, which ultimately produces one or more complex structures consisting of highly organized systems of hyphae. These complex structures are called fruiting bodies. Most basidiomycetes, especially those that produce macroscopic fruiting bodies, reproduce mainly by the production of basidiospores, but others produce one or more types (sometimes as many as five) of asexual spores.

Classification of Subphylum Agaricomycotina

The phylum Basidiomycota is made up of three main subphyla. These are the Agaricomycotina, the Pucciniomycotina, and the Ustilaginomycotina. The Agaricomycotina is the largest subphylum, with about 20,000 species having been described thus far. This number represents well over half of all known members of the entire phylum. Another 8000 species belong to the subphylum Pucciniomycotina. The vast majority of these fungi, commonly referred to as rusts, are parasites of plants, and some of them are of considerable economic importance. The Ustilaginomycotina is the smallest of the three subphyla, with only about 1500 species (Begerow et al. 2006). Many of these fungi (known as smuts) are, like members of the Pucciniomycotina, economically important plant pathogens. Only the Agaricomycotina will be considered in this chapter. Members of the Pucciniomycotina and Ustilaginomycotina will be discussed in Chapter 9.

The wide range of morphological diversity expressed in the Agaricomycotina is difficult to comprehend, and various systems of classification have been proposed in an effort to place apparently similar (and presumably) related forms together in the same group. Earlier systems were based largely upon morphological features that could be observed with the naked eye, but more recent systems, which have used data from molecular studies, have resulted in a considerable rearrangement of taxa. A detailed overview of the currently accepted classification used for the entire subphylum is well beyond the scope of this book. However, a simplified classification for the Agaricomycotina that is rather practical (although not adhering to the most recent taxonomic concepts) is one that recognizes two large assemblages—the hymenomycetes and the gasteromycetes.

In certain older systems of classification, these two assemblages were recognized as taxonomic classes. In brief, the hymenomycetes are characterized by a hymenium (that is, a basidiospore-producing layer) that is exposed when mature. As such, the hymenium occurs on a portion (or portions) of the external surface of the fruiting body. Moreover, the individual basidia in the hymenium forcibly discharge their basidiospores. In the gasteromycetes, the basidiospores develop on the inside of the fruiting body and are not forcibly discharged when mature. The hymenomycetes represent the larger of the two assemblages and can be further divided into two different groups—the agaricales and the aphyllophorales. In older taxonomic treatments, these are given formal status as orders.

Although data from recent molecular studies do not support these two orders as natural, it is still convenient to assign a particular fungus to one or the other. In brief, members of the agaricales produce fruiting bodies with gills (described below) or a system of tubes, on or within which the hymenium is located. In contrast, fruiting bodies produced by members of the aphyllophorales either lack these structures or, when they are present, the fruiting body itself is fundamentally different in one or more important respects. For example, the boletes in the agaricales and the polypores in the aphyllophorales both have fruiting bodies that produce basidia and basidiospores in a system of tubes, but the fruiting body of a bolete is fleshy and easily broken apart, while that of a typical polypore is tough, leathery, or woody. The fruiting bodies of some polypores are so hard that they could be used in lieu of a hammer to drive a nail into a board. The various different kinds of fungi that have been assigned to the gasteromycetes and the two groups of hymenomycetes are distinguished from one another largely on the basis of exactly where the hymenium is located and the overall shape of the fruiting body.

Agaricales

Members of the agaricales are certainly more familiar to the average person than any other fungi, because this group includes the macrofungi commonly referred to as mushrooms. It should be pointed out that some mycologists restrict the term "mushroom" to those fungi that produce fruiting bodies with gills beneath the cap, which is the concept followed in this book. Other mycologists use a broader definition of the term to include other kinds of macrofungi, sometimes even including those members of the Ascomycota such as morels and truffles that produce macroscopic fruiting bodies. A more technical term for a mushroom is gilled fungus or agaric. Poisonous mushrooms are sometimes called toadstools, but this term really has no precise meaning.

Morphology of Agarics

A typical agaric consists of a flattened to variously rounded cap (or pileus), borne at the end of a stalk. In many instances, the stalk is located directly below the center of the cap, but in some agarics the stalk is acentric. Stalks are relatively long in some agarics and short to occasionally lacking in others. The latter is often the case for fruiting bodies that occur on elevated substrates (for example, the side of a tree). On the underside of the cap, radiating from the stalk, are thin plates called gills. The gills may be attached directly to the top of the stalk, extend down the stalk some distance, or be free from the stalk. The hymenium is found as a distinct layer on both sides of the gills. The gills are closely spaced in most agarics, and there are often sterile hyphae called cystidia (singular: cystidium) that can project well beyond the basidia and attached basidiospores. In some agarics, the cystidia serve to keep two adjacent gills apart.

When the basidiospores are mature, they are liberated into the spaces between the gills and drop into the air below the cap. Interesting, the individual basidiospores do not simply fall off. Instead, they are literally flicked away from the sterigmata. During the few days when the fruiting body is actively liberating basidiospores, the spores fall at rates as high as several hundred thousand per minute from a modest-sized agaric. Because the stalk elevates the cap above the substrate upon which the fruiting body occurs, the chance of the basidiospores being carried away by air currents is greatly increased. The simply prodigious numbers of spores produced provide an indication of just how long the odds are for a particular example successfully germinating and producing a new mycelium.

Identification of Agarics

The color of the basidiospores is an important feature used for identification of agarics and certain other members of the agaricales. Individual basidiospores are much too small to be observed with the naked eye, but it is easy to make a spore print that will show the color of the spores in mass. This can be done by cutting off the stalk as close as possible to the cap and then placing the cap, gill side down, on a piece of paper. For best results, the cap should be covered with a dish or bowl (or placed inside a plastic bag) to prevent excessive drying and then left undisturbed for at least several hours. When the cap is removed from the paper, the masses of spores that have been liberated from the gills and deposited on the paper are apparent as a spore print. The range in spore color extends from pure white to black, with a variety of colors and shades in between.

A number of color groups have been recognized by mycologists and used as a first step in the identification of an unknown agaric. For example, in his very popular field guide *Mushrooms of North America*, Orson Miller (1973) began his treatment of the agarics by dividing them into with six spore color groups. These were (1) white to pale yellow (or pale green in one exceptional example), (2) salmon to pink, (3) black to smoky gray, (4) purple-brown to chocolate-brown, (5) clay colored, and (6) bright yellow-brown, rusty brown, cinnamon-brown to clay-brown. Most of these are fairly easy to recognize by even a novice, but others can be distinguished without difficulty only after one acquires a little experience.

The fruiting bodies of some agarics consist of more than just a cap and a stalk. In some examples, when the fruiting body first develops, it is completely enclosed within a universal veil. The latter structure serves to protect the immature fruiting body and by extension the basidiospores at a critical stage in their development. However, as the fruiting body increases in size and the stalk elongates, the universal veil ruptures, often leaving behind a large, cuplike remnant (called the volva) around the base of the stalk and smaller remnants (usually referred to as scales) on the upper surface of the cap. The presence or absence of these features, which are best observed in newly expanded fruiting bodies, is exceedingly useful in the identification of some species.

In addition to the universal veil, there is sometimes a covering (called a partial veil) that extends from the margin of the cap to the stalk, thus serving to protect just the developing basidiospores. As the cap expands, the partial veil tears away from the cap margin to expose the gills. After this occurs, the remnants often persist as a ring (or annulus) around the stalk. Whether or not an annulus is present and the form it takes if present are diagnostic features used for distinguishing different taxa.

Other features that are taken into consideration when making an identification include such things as the overall size and shape of the fruiting body; the color of the cap, stalk, and gills; the nature of the surface (for example, whether smooth or with fibers or scales present) of the cap and stalk; whether or not the gills are attached to the top of the stalk; and the color of the cut flesh of the fruiting body. Sometimes, other senses come into play, because the fruiting bodies of certain species have a distinctive odor or taste.

For critical identification to the level of species, features of the spores often have to be considered. This involves the use of a compound microscope, sometimes one with an oil immersion objective and an ocular micrometer. The latter, when properly calibrated, allows the size of spores and other structures to be determined. In some instances, it is useful to know whether or not the spores undergo a color change in the presence of certain chemicals such as iodine. Microscopic features of the spores that are important include size, basic shape (for example, spherical, ovoid, or angular), the type of ornamentation present on the wall of the spore (for example, none, warts, or spines), and the presence or absence of a noticeable germ-pore (a thin area that may have the appearance of a shallow pit) in the wall of the spore.

Agarics with White Spores

In most field guides to the agarics, the color of the spores (generally determined from a spore print) represents an important first step towards identifying a particular specimen. Each of the color groups outlined above encompasses one to several taxonomic families, and it is usually possible to place the specimen in one of these families on the basis of other morphological or ecological considerations. For example, prior to the advent of molecular studies, most mycologists recognized five families of agarics with more or less white spores. These are the Amanitaceae, Lepiotaceae, Hygrophoraceae, Russulaceae, and Tricholomataceae. We now know that the feature (that is, white spores) these families share in common is not an indication that they are closely related. Indeed, some mycologists consider the members of what was once regarded as the Russulaceae sufficiently different from all other agarics to constitute a separate order (Russulales).

Moreover, the traditional concept of the Lepiotaceae as it was presented when I first took classes in mycology has undergone a major revision. This family is now known to be closely related to the Agaricaceae, a group of brown-spored agarics, and many mycologists combine the members of the two families into the single family Agaricaceae. Nevertheless, spore color does provide a convenient and very useful starting point for the identification of an unknown agaric, and some of the more interesting or better known examples in each of the color groups will be discussed here.

Family Amanitaceae
Among the more common and widespread white-spored agarics are members of family Amanitaceae. Some of these probably represent the

most widely known of the non-cultivated (or "wild") agarics, namely, the deadly poisonous species in the important genus *Amanita*.

At least 500 species of *Amanita* are known. All of them produce fruiting bodies with a universal veil (and thus a volva), and most also have an annulus. They are an ecologically very important group of mycorrhizal fungi, and their fruiting bodies are likely to be encountered in virtually any type of temperate or boreal forest. Although best avoided because of the chance of making a potentially fatal mistake in identification, a few species of *Amanita* are edible. Among these are *A. caesarea* (Caesar's mushroom), which was especially prized in ancient Rome, *A. jacksonii* (American Caesar's mushroom), and *A. rubescens* (blusher).

Examples of poisonous species include *Amanita phalloides* (green death cap) and *A. bisporigera* (destroying angel), which are responsible for most mushroom fatalities in North America. *Amanita bisporigera* has a pure white fruiting body and is widely distributed in eastern North America, where it occurs in oak and mixed hardwood forests during the summer and fall (PLATE 48). *Amanita phalloides* is widespread in Europe and appears to have been introduced into North America at some point during the latter half of the twentieth century. The species is now known from scattered localities throughout North America, where it is often associated with oaks. Unfortunately, *A. phalloides* closely resembles several edible species, thus increasing the risk of accidental poisoning.

Amanita muscaria (fly agaric) is easily recognized by its straw-yellow, orange to blood-red cap with numerous white warts that represent fragments of the universal veil. This species is widely distributed throughout most of the Northern Hemisphere and has been introduced to such places as Australia and New Zealand. The toxic effects produced by *A. muscaria* when consumed by humans are, as will be described in Chapter 11, quite different from those associated with *A. phalloides* or *A. bisporigera*.

Family Lepiotaceae

As considered in the traditional sense, the Lepiotaceae is a family of white-spored agarics rather similar in appearance to the Amanitaceae. Evidence from molecular studies suggests that the fungi once considered to comprise the Lepiotaceae are more appropriately placed in the Agaricaceae, a family otherwise made up of brown-spored agarics. There has been some reshuffling of taxa among the genera formerly assigned to the Lepiotaceae. For example, some species in the genus *Lepiota*, which includes most of the more familiar members of the family, have been reassigned to *Macrolepiota*.

Species of *Macrolepiota* can be confused with amanitas, but the former never have a volva and what might sometimes appear to be fragments of a universal veil on the top of the cap in a *Macrolepiota* are scales that are part of the cap itself. Several of the larger species of *Macrolepiota* (for example, *M. procera*) have long been called parasol mushrooms for very obvious reasons. In a particularly large specimen, the cap can have a diameter of more than 8 inches (20 cm).

Chlorophyllum molybdites (green-spored lepiota) is one of the more distinctive species in the Lepiotaceae because of its dull grayish green spores as revealed by a spore print. This spore color is most unusual in fungi.

Family Russulaceae

The white-spored agarics also include members of the family Russulaceae. Two important genera in this family are *Russula* and *Lactarius*. Both are very large, with at least 750 species having been described for *Russula* and more than 400 species for *Lactarius*. Members of the two genera are morphologically similar, and many produce colorful fruiting bodies that are fleshy but typically rather brittle (PLATE 49). There is never any evidence of either a volva or an annulus, and the spores are ornamented with warts and ridges. This group of mycorrhizal fungi is common and widespread in both coniferous and broadleaf forests throughout the world.

Most species of *Lactarius* produce a watery or milklike latex if the fruiting body is cut or broken when still fresh. For this reason, they are often referred to as milk mushrooms or milk caps. Some species in both *Russula* and *Lactarius* are edible, but others can cause gastrointestinal problems if consumed. Specimens in both genera are often difficult to identify to species because there are so many morphologically similar forms. For *Lactarius*, the color of the fresh latex and any color changes that occur after the latex has been exposed to air are often important diagnostic features.

Family Hygrophoraceae

The fruiting bodies produced by members of the Hygrophoraceae tend to be quite different in appearance from those of the other white-spored agarics described thus far. In this family, the fruiting bodies range from small, delicate, brightly colored examples to those that are more robust, fleshy, and have more subdued colors. Their most important distinguishing feature is the presence of thick, waxy gills, which actually do feel like candle wax when rubbed between the fingers. Often, the cap also has a waxy appearance, which accounts for the common name "waxy caps" given to the Hygrophoraceae as a group.

There are two important genera, *Hygrocybe* and *Hygrophorus*. Species of *Hygrocybe* are saprotrophs associated with dead organic matter and have small and brightly colored fruiting bodies (PLATE 50). In contrast, the genus *Hygrophorus* comprises species that form mycorrhizal associations with forest trees.

Family Tricholomataceae

The last of the traditional families of white-spored agarics is the Tricholomataceae. Until data from molecular studies became available, this very large and diverse family included numerous taxa now placed in other families, two of which (the Marasmiaceae and the Mycenaceae) will be discussed below.

As currently recognized by mycologists, the Tricholomataceae encompasses several genera that are frequently encountered and often relatively easy to identify. The most prominent of these are *Armillaria* (which some mycologists place in a separate family, the Physalacriaceae), *Clitocybe*, *Laccaria* (sometimes placed in the separate family Hydnangiaceae), *Lepista*, and *Tricholoma*. *Laccaria* and *Tricholoma* are mycorrhizal fungi, *Armillaria* is parasite of forest trees but can also occur as a saprotroph on woody debris, while *Clitocybe* and *Lepista* are saprotrophs on forest floor litter.

All five genera contain edible species. Prominent among these are *Armillaria mellea* (honey mushroom), *Clitocybe odora* (anise-scented clitocybe), *Laccaria laccata* (common laccaria), *Lepista nuda* (wood blewit), and *Tricholoma matsutake* (matsutake or pine mushroom). The latter is a highly valued edible fungus in China and Japan.

The taxon commonly identified as *Armillaria mellea* actually consists of several species that are morphologically almost identical and have not always been recognized as distinct entities. This situation is not uncommon, and in a number of other instances in the agarics a species complex is known or suspected.

Family Marasmiaceae

Species of *Marasmius* are the most frequently encountered members of the Marasmiaceae and occur as saprotrophs on forest floor litter. The fruiting bodies produced by these fungi are typically small and rather tough, and have somewhat flattened caps with widely spaced gills and long (hairlike) stalks that are often quite flexible. The cap shrivels up as it dries out, but unlike the case for most other agarics, it revives upon being rewetted. Several of the more common species, including *M. rotula* (pinwheel mushroom) and *M. siccus* (orange pinwheel marasmius), often can be found in groups on dead leaves (PLATE 51).

Family Mycenaceae

The most important genus in the Mycenaceae is *Mycena*. The typically small fruiting bodies of species of *Mycena* might be confused with those of *Marasmius*, but in *Mycena* the fruiting bodies are fragile instead of being tough, the cap is conical or bell-shaped, and the gills are not noticeably widely spaced. Some species of *Mycena* (for example, *M. haematopus*) produce a watery exudate when the stalk is broken (PLATE 52). As a group, mycenas are more colorful than species of *Marasmius*.

Agarics with Salmon to Pink Spores

The second spore color group (salmon to pink) consists of just two families, the Pluteaceae and the Entolomataceae. Members of the Entolomataceae are mostly mycorrhizal and occur on the ground. Their fruiting bodies have gills that are attached to the top of the stalk, and the spores are unusual in that they are extremely angular or longitudinally ridged. The most commonly encountered genera are *Entoloma*, *Nolanea*, and *Leptonia*.

In contrast, members of the Pluteaceae produce fruiting bodies with gills that are free from the stalk and spores that are usually ellipsoidal and smooth. These fungi are saprotrophs on decaying wood. *Pluteus cervinus* (deer mushroom) is a common and widespread representative of this family (PLATE 53). This species occurs on decaying wood in eastern North America throughout the entire growing season. It is edible but remains in fresh condition for only a short time.

Agarics with Black to Smoky Gray Spores

The agarics with black to smoky gray spores include two families. One of these (the Coprinaceae) is of particular interest. As described earlier for the Tricholomataceae, the traditional concept of the Coprinaceae has undergone some major changes as a result of data from molecular studies. In fact, several species previously assigned to the genus *Coprinus*, including the species (*C. comatus*) from which the fundamental concept of the family Coprinaceae was derived, are better placed in other families. Moreover, a number of other species have been moved from *Coprinus* to such "new" genera as *Parasola*, *Coprinellus*, and *Coprinopsis*.

Nevertheless, as circumscribed in the older sense, *Coprinus* includes those agarics that have long been known as ink caps because they have gills that are deliquescent (literally turning into liquid). As a group, these fungi occur as saprotrophs on soil, dung, and wood. Perhaps the best-known species is *C. comatus* (shaggy mane), which is often found growing out of disturbed soil at the edge of roads and trails (PLATE 54). If collected while still fresh, *C. comatus* is edible.

The more commonly encountered species of *Coprinopsis* include *C. atramentaria* (also listed as *Coprinus atramentarius* in many field guides), which occurs in grassy areas such as lawns and on various types of woody debris. This fungus is edible but only when an alcoholic beverage is not part of the same meal and/or has not been consumed several hours before or after the meal in question. When *Coprinopsis atramentaria* is consumed along with alcohol, the fungus usually produces such unpleasant symptoms as nausea and vomiting.

Agarics with Purple-Brown to Chocolate-Brown Spores

Two families (Strophariaceae and Agaricaceae) make up the assemblage of fungi with purple-brown to chocolate-brown spores. The Strophariaceae is a family of saprotrophic fungi found on soil, dung, or wood. The best-known genus is *Psilocybe*, some species of which are hallucinogenic. When someone mentions "little brown mushroom" in a conversation, there is an excellent chance that what they are talking about is a member of the genus *Psilocybe*.

In the traditional sense, the most important genus in the Agaricaceae is *Agaricus*, which includes both the mushroom of commerce (*A.*

bisporus) as well as several other fungi that produce relatively large, fleshy fruiting bodies. All can be distinguished by their chocolate-brown spores. Although *A. bisporus* is edible, some other species in the genus are not. *Agaricus bitorquis* (spring agaricus), one of the edible species, has another distinction. The fruiting bodies of this fungus, which commonly occurs in urban areas, have been known to force their way up through very hard soil or even asphalt. For this reason *A. bitorquis* is sometimes referred to as the pavement mushroom.

Agarics with Clay-colored Spores

In the categories for spore color used in Orson Miller's book, only a single family (the Paxillaceae) is listed as having clay-colored spores. One of the more surprising things to emerge from the molecular studies of agarics is that the members of this family are actually part of the assemblage of fungi recognized as boletes. As will be described later in this chapter, the boletes lack gills. The fruiting bodies of *Paxillus*, the genus in the Paxillaceae most likely to be encountered, clearly have gills. Members of this family are not closely related to the other families discussed thus far, and their gill-like structures evolved independently of the gills found in the "true" agarics.

Agarics with Bright Yellow-Brown to Clay-Brown Spores

The last of the six spore color groups (bright yellow brown to clay brown) encompasses several families, one of which (the Cortinariaceae) includes a genus of special significance. In most major groups of organisms, at least one very large genus provides a challenge to the taxonomists who study it and an almost impossible dilemma to anyone who does not. Such is the case for the genera *Carex* and *Senecio* (each with more than 1000 species) among the vascular plants and the genus *Eleutherodactylus* (more than 700 species) for the amphibians. For the agarics, this is the genus *Cortinarius* in the Cortinariaceae.

Cortinarius is estimated to include at least 2000 species. The genus itself is fairly easy to recognize because of its unusual partial veil, which consists of numerous individual filaments and has the general appearance of a spider web. This type of partial veil is called a cortina. Species of *Cortinarius* form ectomycorrhizae with forest trees, especially conifers. In some years, their fruiting bodies are so abundant in the white spruce (*Picea glauca*) forests of central Alaska that it is difficult to walk across the forest floor without stepping on them.

Other genera in the Cortinariaceae include *Galerina*, *Hebeloma*, *Inocybe*, and *Rozites*. Most of these are mycorrhizal fungi, but a few occur as saprotrophs on decaying wood. With the exception of *Rozites caperatus* (the gypsy), which is considered edible, these fungi should not be collected for the table. Some are deadly poisonous.

Boletes

The concept of the agaricales used herein includes the boletes (family Boletaceae), a group of fungi somewhat similar in appearance to agarics yet sometimes placed in their own order (the Boletales). The fruiting body of a bolete has a cap and a stalk, but the basidiospores are produced within a system of tubes instead of on gills. As viewed from below, the lower surface looks as if it is covered with thousands of little holes. Each hole (or pore) is the end of one of the tiny tubes. The size and arrangement of the pores are features useful in the identification of the different kinds of boletes. Spore color can be determined by making a spore print in the same manner as described for the agarics.

As a group, the boletes are quite colorful, and some species provide food for deer, squirrels, and other rodents. While many species are edible and thus can be consumed by humans, others have a bad taste and some are actually poisonous.

The largest genus of boletes is *Boletus*, with an estimated 300 species. One of these is *B. edulis* (king bolete), which is a highly prized edible fungus, especially in Europe (PLATE 55). Species of *Boletus* are mycorrhizal, and the same is true for two other commonly encountered genera, *Suillus* and *Leccinum*.

Suillus is a genus of medium-sized, yellow to brown boletes in which the surface of the cap is slimy to the touch. These fungi are found under conifers, especially pines.

Members of the genus *Leccinum* produce relatively large, robust fruiting bodies that are immediately recognizable because of their dry leathery cap and the presence of conspicuous tufts of dark hyphae (called scabers) on the stalk. The fruiting bodies of *L. insigne* (aspen scaber stalk), a species that is mycorrhizal with quaking aspen, can be more than 8 inches (20 cm) tall (PLATE 56).

Aphyllophorales

The aphyllophorales, the second of the two groups of fungi recognized herein as comprising the hymenomycetes, is made up of the polypores, tooth fungi, chanterelles, coral fungi, and the corticioid fungi. The aphyllophorales are morphologically more diverse than the agaricales, and evidence from recent molecular studies has verified the fact—long suspected by some mycologists—that some of the taxa placed in the aphyllophorales actually belong elsewhere. Moreover, the families (for example, the Corticiaceae, Hydnaceae, and Polyporaceae) traditionally recognized within the aphyllophorales are not always made up of assemblages of closely related taxa. In fact, some of the taxa formerly assigned to one of these families have been found to be more closely related to certain members of the agaricales than to other members of the aphyllophorales.

All of the different types of fungi mentioned above as belonging to the aphyllophorales produce fruiting bodies in which the hymenium is exposed during development, but only rarely does the hymenium occur on structures that would be identified as gills. Moreover, when what appear to be gills are present, they are leathery to woody and not fleshy as is the case in the agarics. In some cases, the fruiting bodies are fleshy but in other cases they are tough, leathery, or even woody.

Polypores

Polypores (family Polyporaceae) tend to have very tough, leathery or woody fruiting bodies. Some of these fungi produce a new fruiting body every year, while others produce one that continues to grow year after year. Those in the latter group may reach a considerable size and have visible layers that can be counted in a manner similar to the growth rings that occur in wood.

Polypores are like boletes in that both types of fungi produce their basidiospores in tubes. While the fruiting body of a bolete is soft and fleshy, that of a typical polypore is not. Moreover, in contrast to boletes, which are found on the ground as mycorrhizal fungi, most polypores occur on decaying wood. They are often shelflike and most grow out of tree trunks or decaying logs, although some may appear to grow on soil when the fruiting body arises from a root or piece of buried wood.

Polypores are often referred to as bracket fungi or shelf fungi, because they resemble shelves growing out of the sides of trees.

The fruiting bodies of some polypores are fundamentally more complex than those of most other fungi. They are complex because they can consist of as many three different kinds (or systems) of interwoven hyphae—generative hyphae, skeletal hyphae, and binding hyphae. Generative hyphae are septate and generally thin-walled, skeletal hyphae are thick-walled and non-septate, and binding hyphae are thin-walled, non-septate and highly branched. The skeletal hyphae are largely responsible for making the fruiting body tough or hard.

Any fruiting body in which all three types of hyphae are present is said to be trimitic. In contrast, various other polypores have only two systems present and are referred to as dimitic, whereas a few species produce fruiting bodies made up of only one system and are thus considered monomitic. Two common examples of species that produce dimitic fruiting bodies are *Ganoderma applanatum* and *Laetiporus sulphureus*.

Laetiporus sulphureus (sulfur shelf or chicken of the woods) is one of the more colorful and better known polypores in North America. The orange-yellow fruiting bodies are as much as a foot (30 cm) wide and typically occur in compound clusters on dead stumps, logs, and still-living trees (PLATE 57). This fungus is a prized edible, and the fruiting bodies are so distinctive that they are not easily confused with anything else. One of the common names for *L. sulphureus* is chicken of the woods. To say that something tastes like chicken is a common expression, but in this case it is not entirely inappropriate. One summer when I was carrying out research at the Mountain Lake Biological Station in Virginia, several fruiting bodies of *L. sulphureus* were collected, cut into strips, cooked up, and served as part of the evening meal. It was simply placed out without being identified. This fungus has what is probably best described as a "meaty" taste, and several people actually had the impression that they were eating chicken.

The fruiting bodies of *Ganoderma applanatum* (artist's conk) would never be confused with chicken because they are much too hard and woody to be considered edible. However, just like *Laetiporus sulphureus*, *G. applanatum* is a distinctive polypore. The individual fruiting bodies are dull brown, grayish brown to pale brown and as much as 2 feet (60 cm) wide. The lower surface of the fruiting body is white at first but turns brown when bruised or scratched. The common name of this fungus is derived from the fact that artists sometimes etch drawings on the lower surface of fruiting bodies, and these become more or less permanent. Such drawings often turn up at craft shows.

Trametes versicolor (turkey tail) is yet another widespread, common and colorful polypore (PLATE 58). The fruiting bodies of *T. versicolor* (or *Coriolus versicolor* as it is listed in some field guides) are rarely more that about 3 in (8 cm) wide and less than ¼ inch (5 mm) thick. As such, they are much smaller than those of either *Laetiporus sulphureus* or *Ganoderma applanatum*. The fruiting bodies, which occur in overlapping clusters on dead branches, stumps and logs, are easy to recognize because of their concentric bands of contrasting colors. They resemble the extended tail feathers of a turkey enough to account for the common name of this fungus. The fruiting bodies of *T. versicolor* are trimitic, and they are tough, fibrous, and will bend without breaking. Since they do not decay readily, they have received some use as ornaments (Arora 1979).

Corticioid Fungi

The corticioid fungi (family Corticiaceae) have always represented a problematic group, with some member species included by default since they were not a good fit anywhere else. This exceedingly diverse assemblage of fungi has numerous species worldwide. The vast majority of these are saprotrophs found on decaying wood, but some species can be present in trees that are still living and others are mycorrhizal.

In many instances, corticioid fungi produce thin or crustlike fruiting bodies (that look like a layer of paint) over the surface of the substrate (usually decaying wood) upon which they occur. This growth form is referred to as resupinate (spread out). The hymenium, which often extends over the greater part of the exposed surface of the fruiting body, can be smooth, wrinkled, convoluted, or covered by short toothlike structures.

Molecular studies have clarified the taxonomic placement of many corticioid fungi. For example, it is now known that the family previously recognized as the Corticiaceae—and the source of the common name for this group of fungi—is an artificial grouping of taxa that were not all closely related. Moreover, some taxa that had been assigned to other families (for example, the Polyporaceae and the Stereaceae) did not actually belong there. In a practical (albeit not taxonomic) sense, all of the species that characteristically produce the type of fruiting body described above can be referred to as resupinate fungi (PLATE 59).

Although the corticioid fungi may not be as conspicuous in nature as the polypores, that does not mean that they are any less common. For several summers during the 1990s, I had the opportunity to spend some time in the field with Harold "Hal" Burdsall, a mycologist who worked for U.S. Forest Service. Hal is an expert on resupinate fungi, and the frequency with which he used his hand ax to remove a specimen from some woody substrate underscored the relative abundance of this group of fungi.

Chanterelles

Some of the fungi assigned to the aphyllophorales produce fruiting bodies that might cause them to be mistaken for an agaric. For example, chanterelles (family Cantharellaceae) characteristically produce funnel- or trumpet-shaped fruiting bodies with caps that are slightly to deeply depressed in the center (PLATE 60).

While a chanterelle would seem to resemble an agaric in many respects, the spore-bearing surface of the cap is quite different. Some species have a cap that is nearly smooth underneath, while in others the lower side of the cap has a network of wrinkles or gill-like ridges that also extend down the stalk. The ridges have many forks and cross veins and are always blunt-edged. In contrast, the "true" gills of a typical agaric are sharp-edged.

The common chanterelles are members of the genus *Cantharellus*, and *C. cibarius* (golden chanterelle) and *C. lateritius* (smooth chanterelle) are typical representatives. Chanterelles are highly prized as edible fungi, both to humans and wildlife.

Tooth Fungi

The chanterelles are closely related to the tooth fungi (family Hydnaceae), a group that includes some of the most easily recognized of all fungi. In the tooth fungi, the hymenium is located on toothlike structures that project down from the main portion of the fruiting body. Some tooth fungi resemble agarics by having a stalk and a cap, while others have a shape that is not so easy to describe.

Hydnum repandum (sweet tooth) is probably the best example of the agaric-like group. The fruiting bodies have an orange to buff cap and a whitish yellow to pale yellow stalk, and the toothlike structures are cream-colored. This fungus is both distinctive and edible, which is a combination that works well for anyone collecting for the table.

In contrast, to this rather agaric-like fungus, the large, white fruiting bodies of *Hericium erinaceus* (bearded tooth) are cushion-shaped to somewhat globose and can be as much as a foot (30 cm) across. The fruiting body has the general appearance of a frozen waterfall or the result of a water pipe breaking in the winter, with the rather long toothlike structures looking like icicles. The toothlike structures can be as much as 1¼ inches (3 cm) long (PLATE 61). This fungus is edible, and a single fruiting body can be large enough to serve as a meal for a whole family. The fruiting bodies of *H. americanum* (bear's head tooth fungus) are similar in size to those produced by *H. erinaceus* but differ in having somewhat shorter toothlike structures that occur in clusters at the tips of distinct branches.

Coral Fungi

In the fungi, there is little doubt that appearances can be deceiving, and morphologically similar fungi are not always closely related. This is particularly true for the group of fungi known as coral fungi. These fungi are characterized by fruiting bodies that range from an erect, club-shaped, unbranched structure to one consisting of a series of upright branches arising from a single stalk. Sometimes the upright braches are so repeatedly branched that they form a complex mass. The hymenium covers the upper portion of the fruiting body.

Although usually placed together in the same section of a field guide, the coral fungi actually include representatives of two entirely different groups of fungi. Some belong to a taxonomic assemblage that also includes the chanterelles and tooth fungi, while others are more closely related to the gasteromycetes to be described later in this chapter. The features (for example, the number of basidiospores produced per basidium) that define the two groups are microscopic and thus not possible to observe directly in the field. As such, this is one instance in which the identification of a particular specimen to species often takes place without first knowing the higher taxon (for example, family or order) to which it belongs.

Coral fungi typically occur on the ground, but some species grow from decaying wood. Most species are saprotrophs, but at least a few are almost certainly mycorrhizal. Many coral fungi are brightly colored, with orange, yellow, or red especially common (PLATE 62). Some species are edible, although most are too small to be worth collecting for the table. Others are poisonous, causing acute gastrointestinal problems.

Gasteromycetes

All of the fungi mentioned thus far are members of the agaricales and aphyllophorales. In both groups, the fruiting body produced has a recognizable fertile layer (hymenium) that occurs in an exposed position. This allows the mature basidiospores, when they are forcibly discharged from their basidia, to be dispersed from the fruiting body.

In contrast, the gasteromycetes produce fruiting bodies in which a recognizable hymenium is often lacking, and the mature basidiospores are not forcibly discharged. In fact, the very name of this group of fungi (literally, the "stomach fungi") is derived from the fact that the basidia and basidiospores are found within a mass of fertile hyphae (collectively called the gleba) that is completely enclosed by an outer sterile layer (peridium).

This is not a natural taxonomic assemblage, and some examples are more closely related to certain members of the agaricales or aphyllophorales than to other gasteromycetes. It appears that the evolution of a more or less closed fruiting body in which the basidiospores are not forcibly discharged has happened on a number of occasions. The best-known gasteromycetes are the puffballs, earthstars, bird's nest fungi, and stinkhorns.

Puffballs

The members of the first group, the puffballs, are very familiar fungi to most people. Puffballs produce their spores inside a fruiting body that is usually globose to pear-shaped. The spores are released through an opening (pore) that develops at the top of the fruiting body as a result of the peridium disintegrating or breaking apart.

There are many different species of puffballs, ranging in size from less than an inch (2.5 cm) to well over a foot (30 cm). The largest known example is *Calvatia gigantea* (giant puffball), which sometimes exceeds 20 inches (50 cm) in diameter and can weigh more than 40 pounds (18 kg). Most puffballs are white, tan, or gray when young.

Puffballs are popular food items for a variety of wildlife and are commonly collected for human consumption. No toxins have been reported for the true puffballs, but these fungi are edible only when young and the interior of the fruiting body is completely white and uniform in texture. The age of a puffball can be determined by slicing the fruiting body from top to bottom. As it matures, the white interior becomes yellow and then brown, finally changing to a mass of dark, powdery spores.

Many of the frequently encountered puffballs are members of the genus *Lycoperdon*, and *L. perlatum* (gem-studded puffball) is very common in a variety of habitats (PLATE 63). The small, cone-shaped spines that occur over the upper portion of the fruiting body make this an easy species to recognize. These spines are deciduous, which means that they eventually fall off, leaving behind shallow pits on the peridium. Eventually, when the spores are mature and ready to be released, a pore develops at the apex of the fruiting body. The gem-studded puffball is found on the ground, while the closely related *L. pyriforme* (pear-shaped puffball) occurs on the decaying wood of broadleaf trees, where it often forms dense clusters. Old, weathered fruiting bodies can persist for many months.

Earthballs are sometimes confused with puffballs because the two types of fungi are superficially rather similar, but the two groups are not closely related. In contrast to the puffballs, which have a hymenium, the spores of earthballs develop within small cavities (called locules) that occur throughout the gleba. Moreover, while puffballs are saprotrophs, earthballs are mycorrhizal. *Scleroderma citrinum* (common earthball) is a typical member of this group of fungi (PLATE 64). Although the fruiting body itself is about the same size as a puffball such as *Lycoderdon perlatum*, the peridium of *S. citrinum* is much thicker, and the mature gleba is olive brown to blackish brown. *Scleroderma citrinum* is considered poisonous.

Earthstars

The earthstars (including such genera as *Geastrum* and *Myriostoma*) are very closely related to the puffballs. Indeed the two groups differ in only one important respect. When they are young, earthstars look very much like puffballs, but as the fruiting body matures, the outermost layer of the peridium splits into sections and these curl back to form the distinctive starlike rays, at the same time exposing a spore sac with an opening at the top (PLATE 65).

In temperate regions of the world, earthstars usually appear in late summer or fall. Since the fruiting body is rather tough, it lasts longer than those of most fungi. Old specimens are often found in the spring. In tropical rain forests, earthstars can be exceedingly common at times.

Many earthstars and some puffballs use the force of falling raindrops to help disperse their spores. In these fungi, the spores are ejected in little bursts as falling raindrops strike the outer wall of the spore sac or peridium. It is possible to mimic the action of raindrops by gently "thumping" the upper portion of the fruiting body with a finger. If a portion of the gleba is still present, each thump will generate a "puff" of spores.

Bird's Nest Fungi

Bird's nest fungi are so distinctive that they are not likely to be confused with anything else. The fruiting body resembles a miniature bird's nest containing eggs (PLATE 66). The "eggs" are packets of spores (called peridioles) that are dispersed as a unit from their "nest" (sometimes for a distance of a foot and a half [0.5 m]) by the force of falling raindrops. The outer wall of the peridiole then decays or is eaten by small animals, and the spores within are released.

All species of bird's nest fungi are small, and the funnel- or vase-shaped fruiting body is usually no larger than about ⅜ inch (1 cm) in diameter at the top. Despite its size, this structure is superbly designed to function as a splash cup mechanism. Large raindrops have a diameter of about ⅛ inch (3 mm) and reach a terminal velocity of about 13 to 26 feet (4 to 8 m) per second. When a raindrop strikes the inside of the fruiting body, the displacement of water in the cuplike cavity creates a strong thrust upwards and outwards along the inclined sides of the cup, and the peridioles are forcibly ejected.

In species of the genus *Cyathus*, each peridiole is attached to the inner wall of the cavity by a cord (called a funiculus) of interwoven hyphae, most of which is folded up in the base of the peridiole. When a peridiole is being splashed from the cavity by a raindrop, the force causes the funiculus to be torn away from the wall. At the same time, there is an almost instantaneous expansion of the funiculus, which unwinds rapidly to trail behind the peridiole in flight. At the end of the funiculus there is a small mass of sticky hyphae called a hapteron. The latter is capable of sticking to any object (for example, the stems or leaves of plants) it might encounter. When this happens, the peridiole is checked in flight and jerked backwards.

The funiculus sometimes becomes wrapped around or entangled with the object to which the hapteron is attached. As a consequence, the peridiole is securely attached and remains in place. Presumably, the position of a peridiole some distance above the ground enhances the chances of it being consumed by a large herbivore such as a deer. The spores inside the peridiole survive passage through the digestive tract of the herbivore, germinate, and eventually produce a mycelium and fruiting bodies on deposits of the herbivore's dung.

The fact that bird's nest fungi rely upon raindrops for dispersal of their spore packets has not always been known. Early mycologists put forth several other explanations to account for spore packet dispersal. For example, some thought that small animals carried away the spore packets, while others believed that water filled the nest and then overflowed, carrying some of the packets with it.

The first mycologist to deduce that the fruiting body was actually a splash cup appears to have been George Martin at the University of Iowa, who published a paper on the subject in 1927. A half century later, Harold Brodie (1975) produced what is surely the definitive book on the subject of bird's nest fungi. His book, *The Bird's Nest Fungi*, is both comprehensive and enjoyable to read.

Bird's nest fungi are common on dung, twigs, small branches, and other types of decaying woody debris in nature, but extensive fruitings sometimes occur on the beds of mulch associated with ornamental plants in urban areas. In my experience, college campuses can be particularly productive at times. Because the fruiting bodies of many bird's nest fungi are light to dark brown in color, they are not always easy to spot. Like the earthstars, the fruiting bodies of bird's nest fungi are rather tough and may persist for quite some time.

Stinkhorns

The stinkhorns are a truly extraordinary group of fungi. Most species are instantly recognizable not only because of their distinctive forms but also from their odors. In the stinkhorns, the spores are produced in a foul-smelling mass of slime that forms a covering over the upper portion (called the head) of the fruiting body. The foul odor emitted by the mass of slime attracts flies and other insects. These crawl around in the slime, consuming some spores and picking up others on their feet. When the insects leave, the spores are carried along and thus can be dispersed well beyond the stinkhorn. The dependence of stinkhorns on animals as vectors for their spores is very different from what is the case in most other macrofungi, where spore dispersal is by wind or rain.

Stinkhorn fruiting bodies develop from structures that are oval to round and usually white to light tan in color. These structures resemble eggs. Although firm in texture, the covering (or peridium) over these stinkhorn eggs is flaccid or rubbery and not brittle like that of a real egg. One or several white rhizomorphs (rootlike aggregations of numerous hyphae) connect the base of the peridium to the mycelium of the stinkhorn that occurs within the underlying substrate. Inside the peridium, the actual stinkhorn fruiting body is surrounded by a gelatinous matrix. When the spores are mature, the main axis of the stinkhorn expands and greatly elongates, causing the peridium to rupture. Some people have compared the process with the hatching of an egg. This final extension of the fruiting body places the mass of slime containing the spores some distance above the substrate being decomposed by the stinkhorn, where it can be found by potential insect vectors.

Mutinus and *Phallus* are two of the more common genera of stinkhorns in temperate regions of the Northern Hemisphere. Both are relatively simple in structure, with the fruiting body consisting of a single straight or slightly curved and hollow stalk, with the slime mass at the apex. In *Phallus*, the fruiting body has a clearly defined head, while in *Mutinus* the stalk simply tapers down to an almost pointed apex with no head present. Both typically occur where decaying plant material accumulates, such as near logs or stumps, within compost piles, and in beds of mulch around ornamental plants.

Certain other types of stinkhorns are more complex morphologically, and the fruiting body may have a number of erect arms or rays, which are sometimes connected to form a cagelike structure. Members of the genus *Dictyophora* (a name that translates as "net bearing") have a netlike skirt hanging out from under the head. Although occurring in temperate regions of the world, *Dictyophora* is much more common in the tropics, where fruiting bodies are not uncommon in situations ranging from gardens to the interior of rainforests (PLATE 67).

Several species of stinkhorns seem to have been introduced into areas of the world where they are not native. For example, *Aseroë rubra* (PLATE 2), a rather exotic-looking stinkhorn in which the fruiting body looks somewhat like a starfish (hence the common name starfish stinkhorn), has somehow made its way from Australia to the United States.

Heterobasidiomycetes

A few basidiomycetes are difficult to place in any of the larger taxonomic groups described thus far. In many textbooks, they are simply lumped together and referred to as heterobasidiomycetes. The latter term is derived from the fact that the basidium is either deeply lobed or subdivided into compartments by septa, forming a heterobasidium.

This assemblage includes a wide variety of different fungi that produce fruiting bodies looking like shapeless blobs of jelly, resupinate crusts, or shapes such as "ears" or "tongues." The fruiting bodies tend to be soft or jellylike and most occur as saprotrophs on the bark of trees, on the ground, or on twigs and other coarse woody debris. Some taxa are parasites of living plants, and the anamorphs of others (for example, members of the genus *Rhizoctonia*) form mycorrhizal associations with orchids. The only other characteristic that holds the heterobasidiomycetes together is that they are of relatively little direct significance to humans.

Genus *Tremella*

The term "jelly fungus" is sometimes applied to any fungus with a jellylike fruiting body, but it is especially appropriate for the members of a group of taxa traditionally assigned to the order Tremellales. Although these fungi usually occur on dead wood and have long been considered saprotrophs, it now appears that many species are parasites on other wood-inhabiting fungi. The largest genus is *Tremella*, which consists of about 80 species.

Tremella mesenterica (witches' butter) is a commonly encountered species. The orange to golden-yellow fruiting bodies occur as lobed to convoluted gelatinous masses on the decaying wood of broadleaf trees

(PLATE 68). Like many of the other jelly fungi, this species is most apparent during or immediately following a period of rainy weather. The fruiting bodies are capable of repeatedly drying out and then rehydrating in response to the availability of moisture. The fact that *T. mesenterica* is a parasite of a wood-decaying fungus (usually a member of the genus *Stereum*) that shares the same substrate is rarely apparent, since the host fungus usually exists only as a mycelium.

A similar species, *Tremella fuciformis* (silver ear), is a widely cultivated fungus in China, where it is consumed as food and also has medical uses. The fruiting bodies are gelatinous, white to almost transparent structures consisting of multiple, thin, bladelike lobes.

Genus *Auricularia*

One of the more widely known heterobasidiomycetes is *Auricularia auricula-judae* (wood ear), which forms rubbery, often ear-shaped fruiting bodies that occur in overlapping clusters on dead wood. This is an edible fungus, and a typical fruiting usually provides enough material to be worth collecting. Certain other species in the same genus are widely cultivated in parts of Asia, where they are used in soups and stir-fry dishes. Such is the case for *A. polytricha* in China, which has been cultivated by humans since at least 600 AD, making it the first cultivated fungus for which there is any kind of historical record. Once collected, fruiting bodies of *Auricularia* can be dried, stored in this condition for several months, and then rehydrated just before being added to whatever dish is being prepared.

Genus *Pseudohydnum*

The fruiting bodies of *Pseudohydnum gelatinosum* (jelly tooth) are unusual in that they appear to have features of two very different groups of fungi. The fruiting body itself is jellylike, but its lower surface is covered with short, toothlike white spines on which the hymenium is located. The toothlike spines are similar to those of the true tooth fungi, discussed earlier in this chapter. *Pseudohydnum gelatinosum* is common on the well-decayed wood of conifers, where the fruiting bodies commonly occur in small groups (PLATE 69).

More About Basidiomycetes

Numerous field guides are available that provide descriptions and color images of many of the more common basidiomycetes, including most if not all of the species mentioned in this chapter. In many instances, information is provided on the edibility of a particular species of fungus. In a fairly comprehensive field guide, there are many species recognized as being edible, although only some of these are highly regarded as an item for the table. In North America, there are several hundred species of fungi known to be poisonous. A few of these are deadly poisonous and are responsible for a number of human fatalities each year. Unfortunately, there is no simple test to distinguish edible fungi from those that are poisonous. The best thing to do is to regard any fungus as unsafe to eat unless it can be identified with absolute certainty.

7

Lichens—More Than Just Fungi

Distribution of Lichens   129
Fungal and Algal
    Components   130
Growth Forms of Lichens   132
Crustose Growth Form   132
Foliose Growth Form   133
Fruticose Growth Form   134
Squamulose Growth Form   134
Lichens with a Dual Growth
    Form   135
Lichens as Food for Animals,
    Insects, and Humans   135

Lichens and the
    Environment   137
Reproduction in Lichens   139
Substrates for Lichens   141
Growth Rate of Lichens   142
Role of Lichens in
    Succession   143
Cyanobacteria as
    Photobionts   144
More About Lichens   145

As the title of this chapter indicates, lichens are more than just fungi. What does this mean? The answer relates to the definition of a lichen as a "composite" organism in which a particular kind of fungus is intimately associated with a certain type of eukaryotic green or prokaryotic blue-green alga (or cyanobacterium). The vegetative body (or thallus) that results from the combination of these two different organisms is a truly remarkable structure that bears little resemblance to either of its two component parts. Indeed, a typical lichen is sufficiently different from most fungi that it would be hard to imagine the two being confused.

Interestingly, the true nature of what today we recognize as a lichen was not comprehended until 1866, when the German mycologist Heinrich Anton de Bary first suggested that what had been previously considered as an autonomous organism actually consisted of two different organisms (de Bary 1866). This fact was subsequently recognized by the Swiss botanist Simon Schwendener, although it was not universally accepted at the time (Schwendener 1867).

Distribution of Lichens

Lichens occur in some of the most inhospitable places on the earth and are often the dominant organisms present in some types of habitats in which severe environmental conditions place major constraints on the presence of other living organisms. For example, lichens can be found at elevations above 23,000 feet (7000 m) on some of the highest mountains in the world and only about 250 miles (400 km) from the South Pole in the Antarctic (PLATE 70).

Some examples grow on the surface of bare rocks in deserts where they are subjected to high summer daytime and often freezing nighttime temperatures and extreme desiccation. A few species are known to occur in the almost unimaginable microhabitat represented by the surface layers of porous rocks. These endolithic (literally meaning "within rocks") lichens are not uncommon in certain situations, ranging from the sandstones and granites of the Victoria Land Dry Valleys of Antarctica to sandstone and limestone outcrops in temperate regions of the world.

Although the survival of lichens under some of the earth's most hostile conditions seems remarkable enough, a scientist working with the European Space Agency discovered that lichens can survive unprotected in space. In an experiment led by Leopoldo Sancho from the Universidad Complutense de Madrid, two species of lichens (*Rhizocarpon geographicum* and *Xanthoria elegans*) were sealed in a capsule and launched into space aboard a Russian Soyuz rocket in May 2005. Once in orbit, the capsule was opened and the lichens were exposed directly to the vacuum of space with its widely fluctuating temperatures and cosmic radiation. After fifteen days in space, the lichens were returned to earth, where they were found to have suffered no discernible damage from their time in orbit.

Fungal and Algal Components

Although lichens are indeed composite organisms, most of what one observes as the thallus of a lichen actually consists of material contributed by the fungal component (or mycobiont), with the algal component (or photobiont) usually representing no more than about 5 to 10 percent of the total mass. Different species of lichens are recognized by the fungus component, so the scientific name used for a particular lichen is actually that of the lichenized fungus; the photobiont has its own name.

Of the more than 13,500 species of lichens described to date (Brodo et al. 2001), an overwhelming majority are lichenized ascomycetes. No more than about 20 species are lichenized basidiomycetes. Interestingly, the total number of species of lichens represents about 20 percent of all known fungi and about 40 percent of all ascomycetes.

The taxonomic diversity of the algal component of lichens is much less. Members of only about 40 genera of green algae and 15 genera of cyanobacteria have been identified as lichen photobionts, and just three of these are found in about 90 percent of all lichens (Tschermak-Woess 2000). The unicellular green alga *Trebouxia* is exceedingly common and probably serves as the photobiont in perhaps 75 percent of all lichens.

Although the majority of lichens consist of a single species of fungus and a single species of alga, this is not always the case. At least 500 examples are known in which the thallus of the lichen in question contains both a green alga and a cyanobacterium. The presence of the cyanobacterium actually confers an advantage, since cyanobacteria are capable of nitrogen fixation, and the nitrogen can then be made available to the mycobiont.

One might wonder why such a seemingly unlikely partnership between a fungus and an alga ever evolved. What advantages do the two components gain as a result of their intimate association?

It has long been known that the mycobiont of a particular lichen utilizes organic molecules (often simple carbohydrates such as glucose) resulting from the photosynthesis carried out by the photobiont contained within the same thallus. The photobiont's contribution of organic molecules to the mycobiont can be appreciable. In a study that involved the lichen *Cladonia convoluta*, it was determined that 70 percent of the total yield from photosynthesis was transferred from the photobiont to the mycobiont (Tapper 1981).

In a simple sense, the alga is providing the fungus with food. What the alga receives in return is more difficult to demonstrate but appears to be related mostly to the favorable microenvironment that is created by the thallus itself. The alga is protected from desiccation, shielded from excessive solar radiation, and provided with mineral nutrients that are either extracted by the mycobiont from the substrate upon which it occurs or deposited directly upon the upper surface of the thallus from the atmosphere.

The "partnership" involved in a lichen certainly allows both the alga and the fungus to survive under circumstances that would not support either one living by itself. For example, many common lichens occur on the bare surfaces of rocks in full sunlight. Such situations are too exposed and nutrient-poor to be easily colonized by vascular plants or other organisms. Clearly, this is not the sort of place one would look to find algae (the majority of which are aquatic organisms) or fungi (which would have no way to meet their energy needs).

Many biologists have considered lichens to represent a perfect example of a mutually beneficial association involving two entirely different organisms, usually referred to as a mutualistic symbiosis (or mutualism). However, there is actually more to the association than is readily apparent. In a typical lichen, special hyphae produced by the mycobiont occur among the cells of the photobiont. In some lichens, the fungal hyphae become appressed to the surface of the algal cells, forming a flattened structure called an appressorium. These hyphae often have short peglike extensions (haustoria) that penetrate the individual cells of the photobiont. The translocation of organic molecules from the photobiont to the mycobiont takes place through the fungal hypha-to-algal cell protoplast contact points represented by the tips of the haustoria (singular: haustorium). There appear to be few if any lichens in which this invasion of the photobiont by the mycobiont does not result in the death of some of the algal cells involved. Since the latter are capable of reproducing faster than they are destroyed, this poses no real threat to their partnership. Nevertheless, what the fungus is doing at the expense of its partner is actually little different from what a parasite does to a host. As such, the question might be asked if a lichen is truly an instance of mutualism or does it represent an instance of a special type of controlled parasitism of the photobiont by the mycobiont (Brodo et al. 2001).

Growth Forms of Lichens

Lichens are often classified into four growth forms—crustose, foliose, fruticose, and squamulose. These forms essentially represent different structural arrangements of the fungal hyphae that constitute the bulk of the thallus of any lichen.

Crustose Growth Form

In most lichens, the uppermost surface of the thallus consists of densely aggregated, thick-walled, heavily gelatinized hyphae that form a protective layer known as the upper cortex. This layer can be some shade of gray or greenish gray but more brightly colored pigments (for example, yellow, orange, or brown) are often present. The pigments in the cortex are important to the photobiont, since they reduce the levels of solar radiation reaching the interior of the thallus, and algal cells can be damaged by exposure to too much light.

The lichens that are the most brightly colored are often those found in the most exposed situations, where light levels are the highest. When the thallus of a lichen is wet, such as immediately after a period of rainy weather, the cortex becomes relatively more transparent, thus allowing the green or blue-green color of the photobiont to become much more apparent. As such, a lichen that is a drab gray color when dry may appear bright green or olive when wet.

For lichens in which the thallus has a clearly distinguishable lower surface, there is often a second protective layer. Known as the lower cortex, this layer is somewhat similar in overall structure to the upper cortex but may differ greatly in color and texture. For example, members of the genus *Parmelia* (shield lichens) have an upper cortex that is usually bluish gray to greenish yellow or brown, while the lower cortex is dark brown to black.

Moreover, the lower cortex often has structures that the upper cortex does not. The "algal layer" of the thallus occurs just below the upper cortex and consists of numerous cells of the photobiont component of the lichen enmeshed in a network of thin-walled, loosely packed hyphae (PLATE 71). Between the algal layer and the lower cortex (or below the algal layer if there is no lower cortex) is a layer of loosely packed hyphae called the medulla. This layer often makes up the bulk of the thallus. The medulla is usually white, but in some lichens pigments cause the medulla to appear more colorful.

Although this general plan applies to the majority of lichens one is likely to encounter in nature, there are some exceptions. For example, the simplest crustose lichens lack an organized thallus, and the "body" of the lichen consists of little more than an indeterminate hyphal mass that encompasses groups of algal cells. The true nature of the basidiolichen *Lichenomphalia umbellifera* (lichen mushroom), sometimes listed as *Omphalina ericetorum* (PLATE 72), is not readily apparent unless one examines the very base of the fruiting body, where scattered groups of cells of the alga *Coccomyxa* are enclosed in tiny fungal envelopes consisting of colorless hyphae. These structures resemble a mass of green granules.

Crustose lichens occur on the surface of many different types of substrates (including rocks, wood, soil, glass, roof tiles, pavement, and the bark of trees), where they are usually so closely adherent that they give the impression of being "painted on" the substrate in question (PLATE 73). Any attempt to remove the thallus as a single unit is doomed to failure, so a portion of the substrate with the lichen present must be collected.

Some years ago, I had several opportunities to go collecting with Mason Hale, one of the foremost American lichenologists of the second half of the twentieth century. Hale pointed out that it was actually quite easy to distinguish the lichenologists from other biologists when observed in the field. The lichenologists were the ones with a geologist's pick or a hammer and chisel. Obviously, these would be exceedingly useful for collecting specimens of crustose lichens.

Foliose Growth Form

Most foliose lichens differ from crustose lichens in having a distinct lower cortex in addition to an upper cortex. This is the case for the members of such common and widely distributed genera as *Lobaria* (PLATE 74), *Parmelia*, and *Physcia*. Some foliose lichens lack a lower cortex, and the thallus is attached directly to the substrate upon which it occurs by hyphae from the medulla. More often, a foliose lichen is attached to the substrate by somewhat rootlike structures called rhizines. Rhizines may be unbranched or branched, and some have perpendicular side branches that give them the appearance of miniature bottlebrushes.

Members of one group of foliose lichens (for example, species of *Umbilicaria* and *Dermatocarpon*) are attached to their substrates by a single, relatively short, more or less centrally located peglike holdfast (umbilicus) on the lower cortex. Because this manner of attachment is somewhat reminiscent of an umbilical cord, these lichens are usually referred to as umbilicate lichens (PLATE 75). Lichens that possess this feature are not necessarily closely related, so the umbilicate growth form has apparently evolved on more than one occasion.

Fruticose Growth Form

Unlike crustose lichens and foliose lichens, both of which are characterized by a thallus that is essentially a two-dimensional structure (albeit sometimes hard to discern for some of the simplest crustose lichens), the thallus of a fruticose lichen is a three-dimensional structure. Many different forms exist, ranging from relatively simple hairlike or fingerlike examples to others that are strap-shaped or intricately branched.

When viewed in cross section, the thallus of a fruticose lichen has a radial internal structure, with an outer cortex, a relatively thin algal layer, a medulla, and either a more or less hollow central cavity or a dense central cord. In some fruticose lichens, the thallus is pendent, hanging down from the substrate to which it is attached. Other fruticose lichens have a thallus that is erect, sometimes having the general appearance of a small, intricately branched shrub (PLATE 76).

All fruticose lichens are attached to their substrate at one or a few points. In theory, this makes them much easier to collect than either crustose or foliose lichens, but this is hardly the case for pendent species that occur a considerable distance above the ground on the branches of a tree.

Squamulose Growth Form

Some lichens produce a squamulose or scalelike growth form that is intermediate between the foliose and crustose types described above.

The thallus of these lichens appears to consist of numerous small scales (squamules) that are usually partially free of the substrate upon which they occur. An individual scale is usually no more than $1/16$ to $3/8$ inch (1.5 to 10 mm) in diameter, but these structures can occur in dense patches that are sometimes quite large. Squamulose lichens are particularly common on soil but also can be found on other substrates (for example, the upper surfaces of dead, decorticated logs and stumps).

Lichens with a Dual Growth Form

In most modern textbooks on lichens or in courses within which these organisms are covered, the general categories "microlichens" and "macrolichens" are frequently used. Microlichens include crustose and squamulose lichens, while macrolichens are made up of the foliose and fruticose lichens. Since some species of lichens produce both squamulose and fruticose growth forms, the distinction between members of the two categories is not absolute. The best examples of lichens with a "dual" growth form are members of the very common genus *Cladonia*. Initially, the thallus clearly consists of numerous squamules, but this primary more or less horizontal thallus eventually gives rise to a secondary essentially vertical and sometimes branched thallus.

Lichens as Food for Animals, Insects, and Humans

Although the thallus of a lichen would not, on first inspection, seem like something that any animal would want to eat, lichens actually represent an important source of food for some species. The best-known example is the reindeer and its North American counterpart, the caribou. During the winter, lichens may make up as much as 90 percent of the caribou's diet, and even in summer about 50 percent of its diet consists of lichens. The lichens consumed include both soil-inhabiting forms as well as those that occur above the ground on the trunks and lower branches of trees (Brodo et al. 2001).

The dependence of reindeer on lichens became a matter of particular concern following the explosion of the Chernobyl nuclear reactor in 1986. When the reactor exploded, a considerable amount of radioactive material was released into the atmosphere. Fallout carried by wind and rain contaminated large areas of Eastern Europe and Scandinavia, more than 1000 miles (1600 km) to the northwest. Lichens absorbed some of these radioactive materials, one of which (a radioactive isotope of the metal caesium referred to as Cs-137) was of particular importance. When reindeer consumed these lichens, they became contaminated. In some parts of Scandinavia, where reindeer represent the primary food source and export product, this was a very serious problem. Levels of Cs-137 were so high in reindeer meat that it could not be consumed by humans. The first reports after Chernobyl suggested that this situation might exist for 40 years, based on the half-life of Cs-137. Fortunately, levels of Cs-137 have since fallen to more acceptable levels, but this does not mean that this radioactive isotope is no longer around.

Other larger mammals that sometimes feed upon lichens include the mule deer, whitetail deer, mountain goat, muskox, and moose. None of these is a lichen specialist and thus depends upon lichens to the same extent as reindeer and caribou. In the Pacific Northwest, northern flying squirrels rely upon species of *Bryoria* (brown beard lichens) as their primary winter food source (Maser et al. 1985). Various other small mammals feed upon lichens or use them in nest building (McCune and Geiser 1997). Fragments of lichens (especially foliose lichens) are often used by certain species of birds (for example, hummingbirds) in constructing their nests. In addition to being readily available in nature, lichens used in the construction of nests also provide some degree of camouflage for the nest itself (PLATE 77).

Many insects have been recorded as feeding upon lichens. For some of these, lichens apparently represent their primary food source. Few of these lichenivorous ("lichen feeding") insects have been studied to any real extent, so there is still a lot to be learned about the ecological associations involved (Rawlins 1984). Among the most common insects are the caterpillars of certain species of moths that graze on the lichens occurring on tree trunks. Many of these are members of the Noctuidae (owlet moths), a family that includes many of the drab, robustly built moths that are common around lights at night. Careful observation of the feeding activities of the caterpillars of these moths reveals that they feed only on the upper cortex and the algal layer. The algal-free medulla is typically avoided. As such, it is more a matter of the insect feeding on just the algae than on the entire thallus of the lichen. Various invertebrates other than insects feed upon lichens, with land slugs particularly important in this respect (Lawrey 1983).

Lichens would not seem to be something that most people would care to eat, but there are records of lichens being used as food in many different human societies across the world. More often than not, lichens represent what might be called famine food and are eaten only in times of dire need, but in a few societies lichens are commonly part of the diet. For example, in those societies that depend upon reindeer or caribou as their primary source of food, partially digested lichens from the stomach of an animal that has just been killed are sometimes consumed.

In North America, *Umbilicaria* (rock tripe) is an example of a lichen that has served as an emergency food. When the thallus is moist and pliable, it is not entirely unappetizing. When in the field with a class of students, I have often demonstrated that it is possible to eat a small portion of this lichen, which is rather tough and has a somewhat "earthy" taste.

Some experts believe that the manna of the Bible, the food given by God to the Israelites, was a lichen, possibly *Aspicilia esculenta* or a related species. When dry, it can be lifted from the soil and transported by the wind, thus producing a "rain" of food. In some parts of the Middle East, *A. esculenta* is still gathered as food, usually to be mixed with meal to produce bread.

Lichens and the Environment

In some situations, lichens exert considerable influence on the composition of vegetation. Some ground-dwelling lichens, such as species of *Cladonia* (reindeer lichens, PLATE 76), produce chemicals that leach into the soil and inhibit the germination of seeds and growth of young plants. The net effect is to produce a situation in which a few scattered trees occur in an area otherwise dominated by lichens. The term "lichen glade" has been used for such situations, which are not uncommon at high elevations near the timberline or in high-latitude areas of the Northern Hemisphere. I have observed several good examples along the Dalton Highway well north of Fairbanks in Alaska.

Since lichens typically occur in naturally harsh environments, it might seem surprising that they are particularly sensitive to airborne pollutants, especially sulfur dioxide, which is produced by the burning of fossil fuel. The reasons for this are twofold. First, as will be developed in more detail later in this chapter, the thallus of a lichen is not ephemeral like most of the readily visible structures (for example, fruiting bodies) produced by non-lichenized fungi. The thallus persists not only throughout an entire year but for periods ranging to many years. During this entire time it is exposed to airborne pollutants. Second, lichens obtain most of the mineral nutrients they need to survive from the atmosphere or in rainwater, where these nutrients are present only in very low concentrations. As a result, lichens have evolved remarkably efficient absorption mechanisms.

Unfortunately, lichens cannot distinguish between naturally occurring useful substances and potentially harmful pollutants, which can accumulate in the thallus over time. Eventually, the levels present in the thallus reach the point that they affect either the mycobiont or the photobiont. In general, the photobiont appears to be the more sensitive of the two components. Since there is already a delicate balance between mycobiont and photobiont in the partnership that allows the lichen to exist as an independent biological entity, anything that affects either component can have disastrous consequences. For example, if the pollutants incorporated by the thallus reduce the photosynthetic abilities of the algae, the partnership quickly breaks down and the lichen dies (Brodo et al. 2001).

As long ago as the mid-1800s biologists noticed that lichens were much less common or even completely missing from areas near the centers of large cities. They correctly attributed the observed phenomenon to environmental pollution. Since then, these lichen-free areas (or "lichen deserts") have been documented for cities throughout the world.

Lichens differ in their sensitivity to pollutants, with some species succumbing to relatively low levels of pollution and others being relatively more resistant. As one moves away from a pollution source, the number of lichen species theoretically increases as relatively more pollution-sensitive lichens appear in a progressive sequence that roughly reflects their level of tolerance.

Because the sensitivity of particular species of lichens can be determined, the presence or absence of certain key species can be used as an index of air pollution (Hawksworth and Rose 1970). When levels of pollution drop over time, substrates formerly occupied by lichens can be recolonized. Interestingly, the recolonizing species may not be the same as the original species.

Reproduction in Lichens

Reproduction in lichens is not as straightforward as is the case in most other organisms. Because lichens consist of two (or sometimes three) different organisms, each of these has to be involved in the process of reproduction. Many lichens reproduce asexually, either by simple fragmentation of the thallus or through the production of special types of reproductive propagules that are unique to lichens.

The most commonly encountered reproductive propagules are soredia (singular: soredium). Soredia are small clumps of hyphae enclosing a group of algal cells. These originate in the medulla and algal layer and then erupt through pores or cracks in the upper cortex. Soredia are easily detached from the thallus and can be carried away by wind, water, or animals. Soredia may be produced over the entire upper surface of the thallus or in smaller, clearly delimited areas called soralia (singular: soralium). The shape of individual soralia and their location on the thallus are often useful features in the identification of lichens. Soralia are most common in foliose and fruticose lichens but often occur in crustose lichens as well.

Another type of reproductive propagule produced in about 25 to 30 percent of all foliose and fruticose lichens is a tiny, cylindrical, often fingerlike structure that arises from the upper surface of the thallus. This structure, which is called an isidium (plural: isidia), appears to represent an integral part of the thallus and can occur rather uniformly over the surface of the thallus. As such, an individual isidium is little more than an extension of the thallus that incorporates both mycobiont and photobiont. Most isidia are rather fragile and thus easily broken off. Once this happens, they have the potential of being dispersed for some distance and then being able to establish a new lichen thallus.

In some foliose lichens, small lobelike structures (lobules) develop along the margins of the thallus. Lobules appear to be an integral part of the thallus in the same way as isidia, and the two structures are distinguished largely on basis of their overall shape (dorsiventrally flattened for lobules and cylindrical for isidia). Lobules, which are especially common on species of *Parmelia* and *Peltigera*, have the same function as isidia.

Fragmentation of the thallus (that is, breaking into pieces) is also a highly effective method of asexual reproduction and dispersal found in many lichens.

Since the mycobiont of most lichens is an ascomycete, it is not surprising that when the lichen thallus produces obvious fruiting structures, they resemble those found in various members of this group of fungi (as described in Chapter 4). Apothecia, perithecia, and pseudothecia are the three types frequently encountered in lichens, with apothecia the most common. In lichens, apothecia are usually open disk- or cup-shaped structures. They may occur at the margin of the thallus, elsewhere on the upper surface, or be elevated above the thallus on an erect, stalklike structure referred to as a podetium (plural: podetia).

Podetia are characteristic of species in the genus *Cladonia*. For some of these, the podetia are rather distinctive. In the pixie cup lichens (certain members of the genus *Cladonia*), the main axis of the podetium expands to form one or more miniature cups (PLATE 78). One could almost imagine such cups being used by the mystical creatures associated with the fairy ring depicted in Plate 113. In the British soldier lichens (*Cladonia cristatella*), a bright red apothecium is located at the apex of the podetium (PLATE 79). The common name often used for these lichens is derived from the "red caps" that British soldiers wore during the American Revolutionary War.

In one small but distinctive group of lichens called pin or stubble lichens (including species of *Calicium*), the asci in the apothecium disintegrate at maturity, and the resulting mass of ascospores forms a mazaedium (plural: mazaedia) at the apex of a tiny stalk. Substrates upon which numerous stalks occur resemble beard stubble, hence one of the common names used for the group.

The ascospores produced in the fruiting structures of lichens are tremendously diverse in shape and size. Once liberated from the ascus, they can be dispersed by wind, water, or animals to some new location, where germination may occur. Since only the mycobiont can be derived from an ascospore, this does not necessarily lead to the establishment of a new lichen thallus. In theory, the mycelium produced from a germinated spore can survive on its own for at least a limited period, but as the first step in reconstituting a lichen, the mycelium must come into contact and form an association with its appropriate photobiont.

A few of the species of algae or cyanobacteria found as lichen photobionts (for example, *Nostoc*) are free-living and widespread, but this is not the case for *Trebouxia*, which occurs as the photobiont in perhaps 75 percent of all lichens but is not known as a free-living alga. Presumably, when the fungal mycelium derived from a spore and the cells of a suitable alga living in nature do form an association, they can eventually give rise to an entirely new lichen thallus. The extent to which this occurs in nature is not known, but it must be relatively common since many lichens (including most crustose examples) do not produce asexual propagules. Lichens have been reconstituted from their respective mycobionts and photobionts in the laboratory, but this is usually a laborious and time-consuming process.

Most people have heard of tumbleweeds. These are vascular plants that break free from their roots in autumn and then are driven by the wind to roll around the ground, scattering their seeds far and wide. Such plants have been called vagrant plants. There are many vagrant lichens that do much the same thing. One of these, *Masonhalea richardsonii* (Arctic tumbleweed), is a foliose lichen in which the lobes of the highly branched thallus curl up when dry, forming a loose, ball-like mass (PLATE 80). This mass rolls freely over the ground in areas of Arctic and alpine tundra where the species occurs. It is quite common in some areas of northern Alaska, where the ball-like masses tend to collect in depressions.

Other vagrant lichens (for example, certain species of *Xanthoparmelia*) are common in typically windswept, semi-arid, and sparsely vegetated areas of western North America and other arid and semi-arid regions of the world. In areas of semi-desert in Australia, *Chondropsis semiviridis* is a common vagrant lichen. When moist, the thallus lies more or less flat on the soil surface. When dry, it rolls up into a tight ball and is easily rolled about and dispersed by wind (Seppelt, personal communication).

Substrates for Lichens

Because of their slow growth rates in nature, lichens invariably occur on stable surfaces that are likely to persist for several years or even longer. Many species occur on rocks or as epiphytes (literally "upon plants") on other plants, including the trunks and branches of trees. Others are found on dead wood or soil. There are even some foliicolous ("living on leaves") lichens that typically occur on the older leaves of some flowering plants (PLATE 81). For example, this is often the case for nikau palm in the forests of northern New Zealand.

The bark surfaces of trees vary considerably with respect to texture, moisture-holding capacity, and chemistry. For example, the bark of some trees has a circumneutral pH, while for others the bark is highly acidic. It is not surprising that most lichens exhibit varying degrees of specificity for certain types of trees. Conifers often support an assemblage of lichens that is very different from those associated with broadleaf trees at the same locality.

Rocks exhibit appreciable differences in their physical and chemical properties, and most rock-inhabiting lichens occur only on specific types of rocks. The assemblages of species found on calcareous rocks such as limestone or marble are very different from, for example, those associated with sandstones or granites. When lichens occur on rock surfaces, they slowly decompose the underlying rock by chemically degrading and physically disrupting the minerals of which it is composed. Over time, this contributes to the process of weathering by which rocks are gradually turned into soil. Although this contribution to weathering is usually benign in nature, it can cause serious problems for artificial stone structures, such as gravestones and monuments, where the presence of lichens may result in the inscriptions or features becoming less and less distinct.

Soil-inhabiting lichens, along with microscopic algae and fungi, often play an important role in soil stabilization, particularly in some desert and semidesert ecosystems. In some instances, vascular plants cannot become established except in places where lichen crusts occur on the surface of the soil.

Growth Rate of Lichens

The thallus of a lichen usually grows very slowly. Studies indicate that some of the larger foliose and fruticose lichens can grow at rates up to $3/8$ to $1 1/4$ inches (10 to 30 mm) per year, but these are exceptional. A more typical growth rate figure would be $1/32$ to $1/4$ inch (1 to 6 mm) per year. Crustose lichens grow at a somewhat slower rate, ordinarily no more than about $1/64$ to $1/8$ inch (0.5 to 3 mm) per year. Extremely slow growth rates of around $3/8$ inch (1 cm) in 1000 years have been calculated for thalli of the crustose lichen *Buellia frigida* in the Dry Valleys of Antarctica (Seppelt, personal communication). Considerable differences may exist from year to year.

The basic pattern of growth for a typical foliose or crustose lichen is for the thallus to expand outward uniformly at the margins. The older parts at the center of the thallus may remain active for the life span of the lichen or eventually die. In the latter case, the still growing outer portions persist as a ringlike form (PLATE 82).

Since the rate of growth and the point at which the thallus was established (and thus began growth) can be determined for a particular lichen, it is possible to derive a rough estimate of age for lichens on substrates that remain stable for appreciable periods of time. This dating technique is called lichenometry (Hale 1983). Studies using the technique have obtained ages of more than 1000 years for some Arctic and Antarctic lichens. The technique also has been used with considerable success to date ancient buildings, stone monuments, and glacial moraines.

Role of Lichens in Succession

In nature, lichens must compete with plants for access to sunlight and a substrate upon which to grow. Because of their relatively small size and slow growth rates, they are at a serious disadvantage when competing for suitable growth sites. Consequently, lichens are literally "displaced" into sites that typically will not support the growth of plants.

A major ecophysiological advantage of lichens is that they are poikilohydric. This means that although lichens have little control over the water their thallus gains or loses, they can tolerate irregular and extended periods of severe desiccation. When water is not available in their immediate environment, lichens enter a type of metabolic suspension (known as cryptobiosis) in which most biochemical activity more or less stops. In this cryptobiotic state, lichens can survive the extremes of temperature, radiation, and drought associated with the often harsh environments (for example, the Dry Valleys of Antarctica) in which these organisms can occur.

Cyanobacteria, microscopic algae, and microfungi are usually the first organisms found on newly exposed substrates in nature, but lichens are often the first macroscopic organisms to colonize bare soil and rock surfaces, and the term "nature's pioneers" is frequently used in this context. From an ecological standpoint, the ability of lichens to colonize such surfaces represents the first step in a successional sequence that ultimately can lead to a diverse assemblage of higher plants. The lichens that occur on rock surfaces enhance the weathering process that results from such things as the freeze-thaw cycle, and the lichen thallus intercepts particulate matter blowing or flowing (in water) across the rock surface. Over long periods of time (often measured in centuries), enough soil and soil-like material build up to allow the spores of mosses to germinate and grow. Once established, mosses accelerate the rate of succession by allowing more soil particles and organic debris to accumulate. Eventually, the spore of ferns that can tolerate the high levels of light and the seeds of relatively hardy grasses, sedges, or other plants reach favorable microsites on the rock surface and allow the first higher plants to become established. In time, conditions improve to the point that the seedlings of certain trees and shrubs are able to survive. Once this stage of succession has been achieved, the transition to a diverse assemblage of higher plants is well under way.

During the late 1880s and the early 1900s, the red spruce (*Picea rubens*) forests characteristic of higher elevations in the mountains of West Virginia were subjected to intense logging. In many places, the trees making up these forests literally grew out of a thick organic mat that covered the underlying rocks. When the trees were removed, this organic mat was exposed to full sunlight and eventually dried out. In this condition, the organic mat was subject to becoming tinder for fires, and in some instances this is exactly what happened. Fires sometimes lasted for weeks, and some places the entire organic mat burned away, leaving behind little more than bare rocks. A century or so later, the forest has yet to reclaim these areas, and it is likely that several centuries will pass before this can happen. The fact that numerous lichens have colonized the rocks provides clear evidence that an eventual recovery to pre-logging conditions is possible.

Cyanobacteria as Photobionts

As already mentioned, some lichens have cyanobacteria as photobionts. When such is the case, the cyanobacterium involved is usually *Nostoc*. Members of this genus have the capability of "fixing" nitrogen, which simply means that they can convert atmospheric nitrogen into forms of nitrogen that can be utilized by living organisms. This process is ecologically important because the supply of available nitrogen in nature does not meet the collective demands of all of the organisms present. In other words, nitrogen is often said to be a "limiting factor" in particular ecosystems. Such is the case for mesic old-growth, mid-elevation Douglas fir (*Pseudotsuga menziesii*) forests in Oregon and Washington. *Lobaria oregana*, a nitrogen-fixing lichen, is exceedingly abundant as an epiphyte in these forests, and results obtained in one study (Antoine 2004) indicated that it contributed more than 50 percent of the annual input of nitrogen.

More About Lichens

As a group, lichens are not necessarily the most glamorous of organisms, and only the most conspicuous examples are likely to be noticed by the majority of humans. Nevertheless, it would be a mistake to dismiss them as insignificant. As described in this chapter, lichens are ecologically important in several different ways, and they represent the dominant organisms in some terrestrial habitats. Moreover, unlike many other organisms, lichens are present in nature throughout the year.

For anyone with an interest in learning more about these fascinating organisms, the single best source of information is *Lichens of North America* (Brodo et al. 2001). This book is a truly comprehensive treatment of the lichens found in the continental United States and Canada. More than 1500 species are covered, and 800 of these are illustrated with what are often simply stunning color photographs. The book was produced as a joint effort involving Irwin Brodo, a well-known lichenologist, and a husband-and-wife team of photographers, Sylvia and Stephen Sharnoff. Although much too massive to be taken into the field, *Lichens of North America* contains the same type of information (albeit more of it) found in most field guides.

Lichens appear very early in the fossil record and thus have been around for a very long time (Taylor et al. 1997). Because of their amazing ability to survive (and even thrive) under some of the more severe environmental conditions the earth has to offer, one could speculate that lichens would be among the last inhabitants to succumb on a dying earth at some distant point in the future.

8
Slime Molds

Myxomycetes 147	Life Cycle 155
Life Cycle 147	Reproduction in Dictyostelids 156
Reproduction in Myxomycetes 148	Role in Nature and Distribution of Dictyostelids 157
Distribution of Myxomycetes 150	Spore Dispersal 159
Substrate Relationships 151	**Protostelids** 159
Collecting Myxomycetes 152	Life Cycle 160
Food and Other Uses of Myxomycetes 153	Substrate Relationships 160
Myxomyceticolous Fungi 153	Distribution of Protostelids 161
Dictyostelids—The Cellular Slime Molds 155	**Ecological Interactions Among Slime Molds** 161

The slime molds are not true fungi, and they actually have more in common with the paramecium or amoeba that can be observed in a drop of pond water when viewed under the microscope than they do with the true fungi. Nonetheless, slime molds are invariably studied by mycologists and thus warrant inclusion in this book.

Although "slime mold" is not a particularly attractive common name, these organisms exhibit incredibly diverse forms and colors, and some of the fruiting bodies they produce are objects of considerable beauty. There are three distinct groups of slime molds—the myxomycetes (or plasmodial slime molds), the dictyostelids (or cellular slime molds), and the protostelids (or protostelid slime molds). Although molecular evidence suggests that the three groups may share a common ancestry, they are distinctly different in a number of important respects.

Myxomycetes

The myxomycetes are the largest and best known of the slime molds, as well as the only examples likely to be observed directly in nature. Members of the group have been known from their fruiting bodies since at least the middle of the seventeenth century, when the first recognizable description of a myxomycete (the very common species now known as *Lycogala epidendrum*) was provided by the German mycologist Thomas Panckow.

There are suggestions that humans have been aware of myxomycetes much longer. In writings from the ninth century attributed to the Chinese scholar Twang Ching-Shih, there is reference to a certain substance *kwei hi* (literally "demon droppings") that is of a pale yellowish color and grows in shady damp conditions (Alexopoulos and Mims 1979). It is thought by some mycologists that this very likely refers to a myxomycete, perhaps a species such as *Fuligo septica* (PLATE 83), which often achieves a size that makes it readily conspicuous.

Life Cycle

Myxomycetes are characterized by a relatively complicated life cycle that was not understood completely until the late 1880s. In brief, this life cycle consists of two very different trophic (or feeding) stages along with a reproductive stage that bears no resemblance whatsoever to either of the trophic stages.

The first of these two trophic stages consists of uninucleate (single-nucleus) amoeboid cells that may or may not be flagellated. These amoeboid cells, derived from myxomycete spores that have germinated, feed and divide by binary fission to build up large populations in the microhabitats in which they occur. Myxomycetes spend this portion of their life cycle as truly unicellular microorganisms, when their very presence in a given microhabitat can be exceedingly difficult, if not impossible, to determine.

Ultimately, the amoeboid cells give rise to a second trophic stage, which consists of a distinctive multinucleate structure called a plasmodium (plural: plasmodia). This structure is the basis of the common name plasmodial slime mold often used for myxomycetes. The transformation from one trophic stage to the other in the myxomycete life cycle is in most cases the result of fusion between compatible haploid amoeboid cells, which thus function as gametes. The fusion of the two cells produces a diploid zygote that feeds, grows, and undergoes repeated mitotic nuclear divisions to develop into the plasmodium.

Plasmodia have no cell walls and exist as thin masses of protoplasm, which often appear to be streaming in a fanlike shape in the larger, more commonly encountered examples (PLATE 84). Bacteria represent the primary food resource for both trophic stages, but plasmodia also are known to feed upon yeasts, algae, cyanobacteria, and fungal spores (Martin and Alexopoulos 1969; Stephenson and Stempen 1994).

Myxomycete plasmodia usually occur in situations in which they are relatively inconspicuous, but careful examination of the inner surface of dead bark on a fallen log or the lower surface of a piece of coarse woody debris on the ground in a forest, especially after a period of rainy weather, often will turn up an example or two. Most of the plasmodia encountered in nature are relatively small, but some species of myxomycetes are capable of producing a plasmodium that can reach a size of more than 3 feet (1 m) across.

Under adverse conditions, such as drying out of the immediate environment or low temperatures, a plasmodium may convert into a hardened, resistant structure called a sclerotium (plural: sclerotia), which is capable of reforming the plasmodium upon the return of favorable conditions. Moreover, the amoeboid cells can undergo a reversible transformation to dormant structures called microcysts. Both sclerotia and microcysts can remain viable for long periods of time and are probably very important in the continued survival of myxomycetes in some habitats, such as deserts.

Reproduction in Myxomycetes

Ultimately, under suitable conditions, a plasmodium gives rise to one or more fruiting bodies containing spores. The spores of myxomycetes are for most species apparently wind-dispersed and complete the life cycle by germinating to produce the uninucleate amoeboid cells. The fruiting bodies of myxomycetes (PLATE 85) are somewhat suggestive of those produced by some fungi, although they are considerably smaller, usually no more than 1/16 to 1/8 inch (1.5 to 3 mm) tall. Although fruiting bodies can achieve macroscopic dimensions, those of most species tend to be rather inconspicuous or sporadic in their occurrence and thus not always easy to detect in the field. Moreover, fruiting bodies are relatively ephemeral and do not persist in nature for very long.

Identification of myxomycetes is based almost entirely upon features of the fruiting bodies. Fruiting bodies (also sometimes referred to as sporophores or sporocarps) occur in four generally distinguishable forms or types, although there are some species that regularly produce what appears to be a combination of two types.

The most common type of fruiting body is the sporangium (plural: sporangia), which may be sessile or stalked, with wide variations in color and shape (PLATE 86). The actual spore-containing part of the sporangium (as opposed to the entire structure, which also includes a stalk in those forms characterized by this feature) is referred to as a sporotheca. Sporangia usually occur in groups, because they are derived from separate portions of the same plasmodium.

A second type of fruiting body, an aethalium (plural: aethalia), is a cushion-shaped structure without a stalk. Aethalia are presumed to be masses of completely fused sporangia and are relatively large, sometimes exceeding a couple of inches (5 cm) in extent. The largest known example, produced by *Brefeldia maxima*, can be more than 1½ feet (0.5 m) across.

A third type of fruiting body is the pseudoaethalium (plural: pseudoaethalia), literally a "false aethalium." This type of fruiting body, which is comparatively uncommon, is composed of sporangia closely crowded together. Pseudoaethalia are usually stalkless, although a few examples are stalked.

The fourth type of fruiting body is called a plasmodiocarp. Almost always stalkless, plasmodiocarps take the form of the main veins of the plasmodium from which they were derived. The fruiting bodies of *Hemitrichia serpula*, a species that can be exceedingly common in moist tropical forests, are classic examples of plasmodiocarps (PLATE 87).

A typical fruiting body consists of up to six major parts: hypothallus, stalk, columella, peridium, capillitium, and spores. Not all of these parts are present in all types of fruiting bodies. The hypothallus is a remnant of the plasmodium sometimes found at the base of a fruiting body. The stalk (also called a stipe) is the structure that lifts the sporotheca above the substrate. As already noted, some fruiting bodies lack a stalk.

The peridium is a covering over the outside of the sporotheca that encloses the actual mass of spores. It may or may not be evident in a mature fruiting body. The peridium may split open along clearly discernible lines of dehiscence, as a preformed lid, or in an irregular pattern. In an aethalium, the relatively thick covering over the spore mass is referred to as a cortex rather than a peridium. The columella is an extension of the stalk into the sporotheca, although it may not resemble the stalk.

The capillitium consists of threadlike elements within the spore mass of a fruiting body. Many species of myxomycetes have a capillitium, either as a single connected network or as many free elements called elaters. The elements of the capillitium may be smooth, sculptured, or spiny, or they may appear to consist of several interwoven strands. Some elements may be elastic, allowing for expansion when the peridium opens, while other types are hygroscopic and capable of dispersing spores by a twisting motion.

Spores of myxomycetes are quite small and range in size from slightly less than 0.002 to 0.006 inch (5 to 15 µm). Nearly all of them appear to be round, and most are ornamented to some degree. Spore size and also color are very important in identification. Spores can be dark or light to brightly colored.

Distribution of Myxomycetes

There are approximately 875 recognized species of myxomycetes, and these have been placed in six different taxonomic orders: Ceratiomyxales, Echinosteliales, Liceales, Physarales, Stemonitales, and Trichiales. Members of the Ceratiomyxales (PLATE 88) are distinctly different from members of the other orders, and many modern mycologists have removed these organisms from the myxomycetes and reassigned them to another group of slime molds, the protostelids.

Most species of myxomycetes are probably cosmopolitan, and at least some species apparently occur in any terrestrial ecosystem with plants (and thus plant detritus) present. A few species appear to be confined to the tropics or subtropics, and others have been collected only in temperate regions. Compared to most other organisms, myxomycetes are widespread, with the same species likely to be encountered in any habitat on the earth where the environmental conditions suitable for its growth and development apparently exist.

For one ecological group of myxomycetes, such environmental conditions apparently occur only at the margins of melting snowbanks in montane areas of the world, such as those found in the Rocky Mountains of western North America, the Alps of Europe, and the Snowy Mountains of southeastern Australia. In such areas, where there is enough snowfall during the winter to produce accumulations of snow sufficiently large to persist until late spring and early summer, the species associated with this rather special and very limited microhabitat are common.

These "snowbank" or "snowmelt" myxomycetes produce fruiting bodies only during the relatively brief period of time when the snowbank melts back. During the remainder of the summer, the species of myxomycetes found in these montane areas are the same as those one would expect to encounter at lower elevations in the same region. Fruitings of snowbank myxomycetes usually occur within a yard or two (1 to 2 m) of the margin of a snowbank and typically are found on ground litter or the stems and leaves of low-growing shrubs and other plants. Interestingly, the fruiting bodies of snowbank species tend to be more robust than those of many other myxomycetes, and the majority of species in such genera as *Lamproderma* and *Lepidoderma* are predominantly associated with snowbanks.

Substrate Relationships

Although the ability of a plasmodium to migrate some distance from the substrate upon or within which it developed has the potential of obscuring myxomycete-substrate relationships, fruiting bodies of particular species of myxomycetes tend to be rather consistently associated with certain types of substrates. For example, some species almost always occur on the decaying wood or bark of coarse woody debris, whereas others are more often found on dead leaves and other plant debris and only rarely occur on wood or bark. In addition to these substrates, myxomycetes also are known to occur on the bark surface of living trees, on the dung of herbivorous animals, in soil, and on aerial litter (that is, dead but still attached leaves and other plant parts).

The myxomycetes associated with decaying wood are the best known, because the species typically occurring on this substrate tend to be among those characteristically producing fruiting bodies of sufficient size to be detected easily in the field. Many of the more common and widely known myxomycetes, including various species of *Arcyria*, *Lycogala*, *Stemonitis*, and *Trichia*, are almost invariably associated with decaying wood. In contrast, most species of such genera as *Diderma* and *Didymium* are found almost exclusively on dead leaves and other types of non-woody plant debris.

In a temperate forest, virtually all of the fruitings large enough to be readily observed in the field are associated with dead plant material (mostly forest floor litter and coarse woody debris) in contact with the ground. In a moist tropical forest, this is not the case, and many fruitings are found well above the ground on various types of aerial litter, including dead portions of vascular epiphytes and lianas (Stephenson and Stempen 1994).

Collecting Myxomycetes

Throughout temperate regions of the world, the fruiting bodies of myxomycetes tend to be relatively abundant only during summer and early autumn. During the remainder of the year, fruiting bodies are encountered only occasionally, and most of these are old specimens that have persisted in nature for weeks or even months.

A simple technique can be used to obtain both plasmodia and fruiting bodies of myxomycetes in the home or laboratory. This technique, which makes use of moist chamber cultures, simply involves placing small pieces of tree bark or other types of dead plant material (for example, dead leaves from the forest floor) on a piece of filter paper (or absorbent paper towel trimmed to the appropriate size) in some type of shallow container that has a lid. Plastic disposable Petri dishes are ideal containers, but it is also possible to use such things as empty plastic butter dishes. After the bottom of the container has been more or less covered with the pieces of the dead plant material, enough water (distilled water if available and tap water if not) is added to cover this material, the lid is placed on the container, and the latter left undisturbed overnight. The following day, the lid is removed, most of the water is poured off, and the lid replaced.

At this point, the container can be recognized as moist chamber culture, and this is checked on a regular basis for a period of several weeks or sometimes even longer. Since a culture tends to slowly dry out, small amounts of water are added from time to time to maintain moist conditions. Most cultures prepared in this manner will yield myxomycetes, and it is not unusual to have the fruiting bodies of several different species appear in the same culture (Stephenson and Stempen 1994). *Echinostelium minutum* (PLATE 86) is an example of a species that commonly appears in moist chamber cultures.

Food and Other Uses of Myxomycetes

No myxomycete is known to be poisonous, and the fruiting bodies produced by a few species achieve some size. On rare occasions, the early developing aethalia of *Fuligo septica* and *Enteridium lycoperdon*, both of which can be a couple of inches (5 cm) or more across, are collected, fried, and then consumed by humans. The taste has been reported to be somewhat like that of almonds.

More importantly, myxomycetes offer food, shelter, and a breeding place for various species of insects. Beetles are the most studied of the myxomycete-associated insects. Members of the family Leiodidae, particularly species of *Anisotoma* and *Agathidium*, have been collected repeatedly from myxomycete fruiting bodies and plasmodia. Many of these "slime mold" beetles appear to feed on nothing else.

Most slime mold beetles are dark, quite small (usually no more than about ⅛ inch [3 mm] in length), and generally are overlooked in nature. Often, the only evidence of their presence is provided by the "entry holes" they make in larger fruiting bodies. The beetles burrow into such fruiting bodies to feed upon the spore mass and to lay their eggs. In many instances, the fruiting body is destroyed almost completely because of the feeding activities of the beetles. However, close inspection of the beetles reveals that their bodies are "dusted" with spores, and it seems likely that the beetles play a role in dispersing the spores of myxomycetes in nature in somewhat the same manner that birds disperse the seeds of certain plants.

Myxomyceticolous Fungi

I have had the opportunity to collect and study myxomycetes in many different ecological situations throughout the world, and one of the things that has been frustrating on more occasions than I care to remember is the fact that the fruiting bodies of myxomycetes provide an organic substrate that is subject to being colonized by fungi. This was very apparent in 1987 when I spent three months in India during the monsoon season, but the same could be said for numerous collecting trips made to tropical rain forests in such places as Costa Rica and Puerto Rico. Any fruiting bodies that were more than a day or two old invariably became moldy as a result of being colonized by a fungus.

These so-called myxomyceticolous fungi tend to smother a given fruiting body, producing a mycelium over its entire surface. Fungal hyphae rapidly penetrate the spore mass of the host, where they invade the individual spores. Ultimately, most if not all of the spores present in the fruiting body are adversely affected by the fungus. Whether myxomyceticolous fungi should be considered as saprotrophs or parasites is somewhat problematic, but the end result is that the spores are rendered non-viable and are never liberated from the fruiting body.

Assessing the ecological impact that these fungi have on myxomycetes is difficult, but in the type of situation described above, it is likely to be considerable. Some mycologists who work with myxomycetes have hypothesized that one of the limiting factors for myxomycetes in tropical forests, where these organisms are found to be less common than might have been anticipated, is the constant high humidity, which promotes the colonization of their fruiting bodies by fungi.

Some years ago, I collaborated with the late Clark Rogerson, a mycologist at the New York Botanical Garden, on a paper in which we recorded the species of both myxomyceticolous fungi and their myxomycete hosts for a large series of collections of "moldy" myxomycetes from several different localities, including Alaska, the eastern United States, New Zealand, and northwestern India (Rogerson and Stephenson 1993). Thirty-five different taxa were represented among the collections considered in this study, although some of these involved the anamorph and teleomorph states of the same fungus. *Verticillium rexianum* was the most commonly encountered myxomyceticolous fungus and occurred on a wide range of hosts. In contrast, some taxa displayed varying degrees of host specificity. *Polycephalomyces tomentosus* almost always occurred on representatives of the order Trichiales, particularly on species of *Arcyria*, *Metatrichia*, and *Trichia*. This fungus is easily recognized by the rod-shaped synnemata that radiate outward from the fruiting body (PLATE 89). The most host-specific myxomyceticolous fungus was *Nectriopsis violacea*, which appeared to be restricted to a single species, *Fuligo septica*. Interestingly, this species produces perithecia that are violet to purple in color, and fruitings of *F. septica* with the fungus present were once considered by some mycologists to represent a separate species, *F. violacea*.

Dictyostelids—The Cellular Slime Molds

The dictyostelids were originally considered fungi. In fact, these organisms are easily mistaken for some of the microfungi that commonly occur as contaminants in laboratory cultures. Dictyostelids are essentially microscopic throughout their entire life cycle. Unlike the myxomycetes, only rarely can they be observed directly in nature. Consequently, dictyostelids must be grown under controlled laboratory conditions in order to be studied.

Life Cycle

Since their discovery by the German mycologist Oskar Brefeld in the late nineteenth century, dictyostelids have intrigued biologists. Their life cycle exhibits a curious alternative to the way in which most other creatures on the earth grow, develop, and become multicellular, with different specialized tissues produced as a result of the process. Most plants and animals begin life as a single cell (called a zygote) that is the product of the fusion of an egg cell and sperm cell. Shortly after the two cells fuse (through a process termed "fertilization"), the zygote divides into two cells that stick together. These cells soon divide again to produce a cluster of four cells that in turn divide, and so on. Within hours or days (depending upon the particular plant or animal), clusters of dozens to thousands of cells form an embryo. Specialized cells begin to take form, and the basic shape of the body of the organism begins to become apparent (Raper 1984).

Dictyostelids approach development differently. Like fungi and myxomycetes, they produce spores as reproductive structures. When a spore germinates, it releases a single amoeboid cell that begins to engulf and digest bacteria in soil and decaying plant debris, the usual habitats for dictyostelids. When the amoeboid cell divides, the two cells separate and become completely independent of each other, with each cell continuing to feed and undergo additional divisions for a number of hours or days. Only after the growing population of amoeboid cells depletes the local supply of bacteria is there any indication that a multicellular structure will be produced.

Reproduction in Dictyostelids

In response to the production of chemical attractants, thousands of amoeboid cells that have been operating as individual single-celled organisms begin to move, either singly or in streaming masses, to form multicellular clumps, or aggregations. Shortly thereafter, one or more cigar-shaped structures called pseudoplasmodia emerge from each aggregation. A pseudoplasmodium is a unified collection of thousands of independent amoeboid cells. The cells remain distinct in the pseudoplasmodium but no longer act independently. Instead, they cooperate as parts of a multicellular entity. Remarkably, when amoeboid cells of two or more different species of dictyostelids are grown together, the amoeboid cells of the different species can recognize each other, so that the cells of any one aggregation tend to be all one species rather than a mixture.

Either immediately or, in some species, after the entire structure has migrated a short distance towards a light source, cells of the pseudoplasmodium begin to display different patterns of specialization. Cells that happen to have been positioned near the anterior end of the moving "cigar" begin to secrete a wall of cellulose. These cells bind together to form a slender stalk that grows upward from the surface of the substrate upon which the pseudoplasmodium occurs. Other cells, those nearer the posterior end of the pseudoplasmodium, are lifted off the surface on the end of the extending stalk. These cells begin to become spores. Only the spores live on and produce another generation of amoeboid cells to feed upon soil bacteria. The cells that produced the stalk in order to elevate the spore cluster above the substrate eventually die, dry up, and decay.

This sequence of events constitutes an asexual method of reproduction, and it appears that dictyostelids reproduce asexually most of the time, at least under laboratory conditions. All of the cells that originate from the same spore are basically genetically identical to one another and collectively represent a clone. As is the case for asexual reproduction in other life forms, finding a "mate" is not necessary to perpetuate the species. If amoeboid cells are equipped with the genetic characteristics necessary to survive long enough to produce spores, the same gene combinations will be passed faithfully to all offspring, thus providing the same qualities for survival.

A method of sexual reproduction, with its potential of introducing genetic variability, also seems to exist in dictyostelids. Occasionally in laboratory cultures, a number of large, thick-walled cells are found that are quite different from spores or encysted amoeboid cells. These giant cells (called macrocysts) appear to form when several amoeboid cells (sometimes described as being of compatible "mating types") fuse together and rearrange their genetic libraries and those of other amoeboid cells that may be engulfed. When macrocysts germinate, the amoeboid cells that emerge seem to have different combinations of genetic information than the cells that initially formed the macrocysts. This mixing of genetic information, along with the genetic changes resulting from mutations, provides dictyostelids with an ability to cope with changing environments.

The actual fruiting body produced by a dictyostelid typically consists of an elongated, erect to semi-erect stalk (called a sorophore) that bears a mass of spores (sorus) at the tip (PLATE 90). In some species, the sorophore is branched, with each branch terminated by a sorus, whereas in other species the sorophores are unbranched, although occasional sparse, irregular branches may be present. The dimensions and branching patterns of dictyostelids vary greatly in different species. Members of the genus *Polysphondylium* are characterized by lateral branches that occur in regularly spaced whorls (PLATE 91), which bear at least a token resemblance to the whorled branches found in the plants called horsetails (*Equisetum*).

As a group, the dictyostelids are not especially colorful, and the fruiting bodies of most species are white to essentially colorless. Some species are strikingly pigmented, ranging from deep purple to bright yellow. Whatever their color, the fruiting bodies of some of the larger dictyostelids possess a subtle beauty that can be appreciated only when these organisms are observed with the use of a stereomicroscope.

Role in Nature and Distribution of Dictyostelids

Most of what is known about dictyostelids has been acquired from studying these organisms in laboratory culture. What about the biology of "wild" dictyostelids in nature? In natural ecosystems, it is quite likely that dictyostelids play a significant role in controlling the size of bacterial populations in soil and decaying litter. Nutrients that are taken up from decaying plants and animals by bacteria are transferred to dictyostelid cells when the latter feed upon these bacteria. The dictyostelids, in turn, become food for soil protozoans, nematode worms, microscopic arthropods such as mites, and other small invertebrate animals. Because of this, dictyostelids play an essential role in patterns of energy flow and nutrient cycles within terrestrial ecosystems.

There are about 130 described species of dictyostelids. These have been assigned to one of three genera—*Dictyostelium*, *Polysphondylium*, and *Acytostelium*. This classification is based solely upon morphology and does not necessary reflect evolutionary relationships. Indeed, molecular studies provide evidence that the three genera do not hold together at all, with some species in two different genera seemingly being more closely related to each other than to species currently assigned to the same genus.

Some species of dictyostelids are found in almost all parts of the world. Two good examples are *Dictyostelium sphaerocephalum* (PLATE 90) and *Polysphondylium pallidum* (PLATE 91). Numbers of species of dictyostelids appear to be highest in the American tropics, which suggests that this region represents a center of evolutionary diversification of the group. More than thirty-five different species have been found in the small area around the Mayan ruins at Tikal in tropical Guatemala. The highest total known from any region in the temperate zone is thirty species for the Great Smoky Mountains National Park in eastern North America (Landolt et al. 2006). In general, numbers of species of dictyostelids decrease with increasing elevation and with increasing latitude. For example, only two species have been reported from extreme northern Alaska.

Some species have restricted habitat associations. One species (*Dictyostelium caveatum*) has been found only in a single cave system in Arkansas. Another species (*D. rosarium*), while recorded from samples collected aboveground in relatively few localities worldwide, also seems to have a particular affinity for the type of conditions found in caves and has been reported from three dozen caves throughout the eastern and central United States from West Virginia to Arkansas (Landolt et al. 2006).

Of the thirty-five species that occur at Tikal, many appear to be restricted to tropical or subtropical regions. *Dictyostelium discoideum*, which is restricted largely to montane forests in the Appalachian Mountains of eastern North America, is the most intensively studied dictyostelid and the one most widely used in research on developmental biology and genetics. Any search for information about dictyostelids will invariably turn up numerous references to this particular species.

Spore Dispersal

Unlike most spore-producing organisms (including myxomycetes), dictyostelids produce spores that do not appear to have the potential for being wind dispersed. Instead, dispersal of dictyostelid spores seems to depend more upon their accidental transport on the body surface or within the digestive tract of some animal. Viable spores of dictyostelids have been recovered from the droppings of a number of animals, including rodents, amphibians, bats, and even migratory birds that travel great distances between their winter and summer homes.

In tropical forests, many living plants and considerable amounts of organic material are found high above the ground in the forest canopy. Dictyostelids have been isolated from the mass of organic material (literally a "canopy soil") found at the bases of epiphytic plants growing on the trunks and branches of trees in these forests (Stephenson and Landolt 1998). It seems likely that they are introduced to such microhabitats by being carried up from the ground by birds, insects, or other animals that move between the forest floor and the canopy above it. Samples of canopy soil collected from a height of as much as 130 feet (40 m) have yielded dictyostelids, and in some tropical forests, they are more abundant in canopy soil than in samples of moist, ground soil collected at the same locality.

Protostelids

The protostelids, the third group of slime molds, were the last to be "discovered" by science. While at Columbia University in New York, Lindsey Olive and his associate Carmen Stoianovitch were attempting to isolate another slime mold from dead grass inflorescences when they noticed the presence of a completely different and apparently unknown organism having minute fruiting bodies, each consisting of a slender tubular stalk bearing a single terminal spore. This organism, later described as *Protostelium mycophaga* (PLATE 92), was the first of the protostelids, a group of slime molds now known to be common and widespread in nature.

The protostelids are the smallest group of slime molds, both in size of the fruiting body and the number of species. There are only about 35 described species, although at least as many other—as yet undescribed—species have been observed by the mycologists who study protostelids.

Life Cycle

With the exception of *Ceratiomyxa*, long regarded as a myxomycete but apparently more closely related to this group, all protostelids produce microscopic fruiting bodies characterized by a delicate, cellular stalk that supports one or a few spores. Trophic stages in the protostelids life cycle are quite diverse, ranging from amoeboid cells (flagellated in some species and without flagella in others) to minute plasmodia that superficially resemble those of myxomycetes. This degree of heterogeneity, which contrasts rather sharply with the situation for both myxomycetes and dictyostelids, suggests that the group of organisms currently classified as protostelids is either extraordinarily diverse or includes examples that might be better placed in some other taxonomic group (Olive 1975).

Substrate Relationships

Because of their very small size, protostelids can be detected only by examining the substrates upon which their fruiting bodies occur with the use of a microscope. The usual method of studying protostelids involves collecting samples of potential substrate material in the field, bringing these back to the laboratory, wetting the samples by placing them in distilled water, and then plating them out on the surface of a weak nutrient agar in a Petri dish. When such laboratory cultures are prepared with samples of various types of dead plant material, protostelids are surprisingly abundant, with several different species often appearing in a single culture. Important substrates for protostelids include dead aerial parts of plants, litter on the ground, the dead outer bark of living trees, bark and wood of coarse woody debris, soil, and the dung of herbivores.

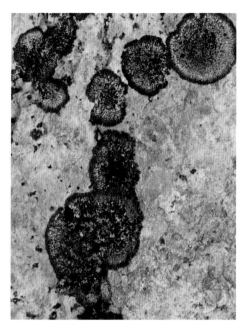

Plate 70. A specimen of the Antarctic lichen *Buellia frigida* in Southern Victoria Land at 77° S latitude. This species is one of the most abundant and widespread Antarctic lichens. Photograph by Rod Seppelt.

Plate 72. The basidiolichen *Lichenomphalia umbellifera* (lichen mushroom) is often not recognized as a lichen. Photograph by Doug Waylett.

Plate 71. The two components of a lichen (fungal hyphae of the mycobiont and the green algal cells of the photobiont) are revealed when a small portion of the thallus is crushed on a glass slide and then viewed under a microscope. Photograph by Carlos Rojas.

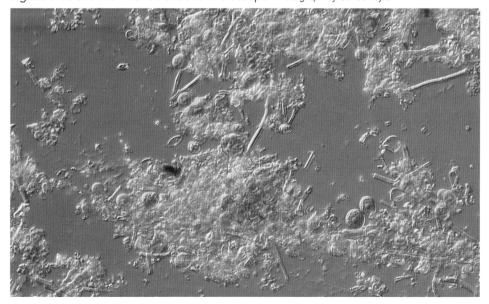

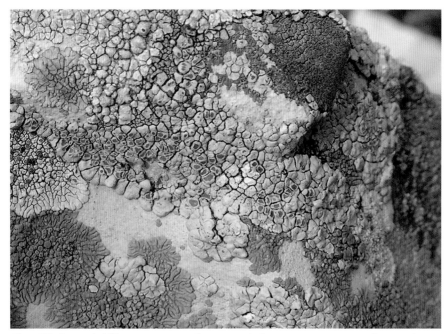

Plate 73. Several different species of crustose lichens may occur together on the same substrate. Photograph by Jason Hollinger.

Plate 74. Many of the more common species of *Lobaria* have a thallus in which the upper surface is marked with a coarse network of ridges. Photograph by Jason Hollinger.

Plate 75. *Dermatocarpon miniatum* (leather lichen). This umbilicate lichen commonly occurs on limestone rocks throughout much of temperate North America. Photograph by Alan Bessette.

Plate 76. Species of *Cladonia* (reindeer lichens) are densely branched. Photograph by Jason Hollinger.

Plate 77. Anna's hummingbird (*Calypte anna*). In this species, the outside of the nest almost invariably has fragments of lichens present. Photograph by Camden Hackworth.

Plate 78. A number of common and widely distributed species of *Cladonia* produce miniature cups and thus can be referred to as pixie-cup lichens. Photograph by Ron Wolf.

Plate 79. The bright red "caps" of British soldier lichens are conspicuous enough to be easily noticed in the field. Photograph by Jason Hollinger.

Plate 80. The Arctic tumbleweed (*Masonhalea richardsonii*) can be common in areas of tundra such as those found in central and northern Alaska. Photograph by Rod Seppelt.

Plate 81. Lichens occurring as epiphylls on the leaves of a plant in a tropical forest. Sometimes there may be as many as two dozen different species of lichens on each leaf. Photograph by Robert Lücking.

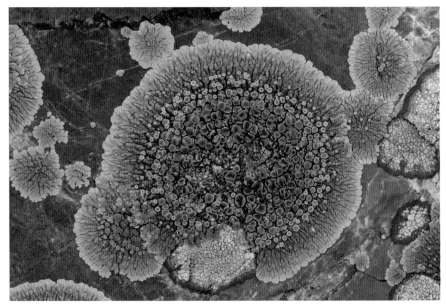

Plate 82. The basic pattern of growth is for a lichen to expand outward from the point at which it was established, which often produces a thallus that has a somewhat rounded outline. The approximate age of the thallus can be estimated if the rate of growth is determined. Photograph by Ron Wolf.

Plate 83. Developing fruiting body of *Fuligo septica*. This species may have been the myxomycete referred to by the Chinese scholar Twang Ching-Shih in the ninth century. Photograph by Bill Roody.

Plate 84. Plasmodium of a myxomycete. Photograph by Randy Darrah.

Plate 85. Fruiting bodies of *Cribraria cancellata* on decaying wood. Photograph by Kim Fleming.

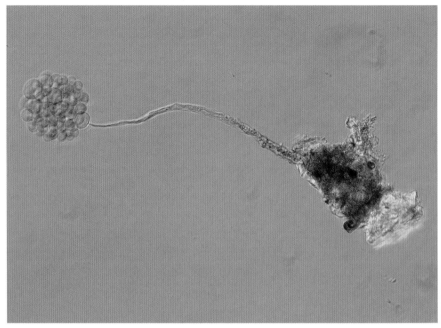

Plate 86. Minute sporangium of *Echinostelium minutum*. Photograph by Carlos Rojas.

Plate 87. *Hemitrichia serpula*. The plasmodiocarps of this species are common in moist tropical forests. This extensive fruiting occurred on a decaying palm frond in a forest in the mountains of central Cuba. Photograph by Randy Darrah.

Plate 88. *Ceratiomyxa sphaerosperma*. Some species of myxomycetes appear to be confined to the tropics, and this seems to be the case for *C. sphaerosperma*. Photograph by Michael Pilkington.

Plate 89. Fruiting bodies of *Metatrichia vesparia* colonized by the myxomyceticolous fungus *Polycephalomyces tomentosus*. Photograph by Harley Barnhart.

Plate 90. *Dictyostelium sphaerocephalum*. Each fruiting body of this dictyostelid consists of a semi-erect stalk bearing a mass of spores at the tip. Photograph by Andy Swanson.

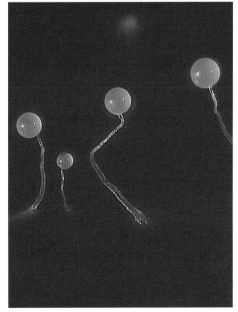

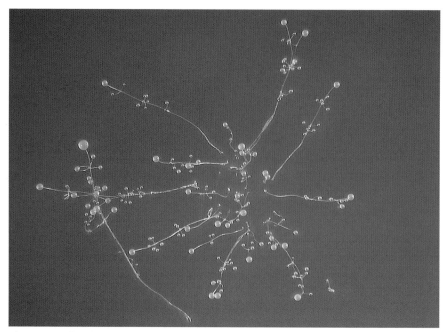

Plate 91. *Polysphondylium pallidum*. The delicate fruiting bodies of members of the genus *Polysphondylium* have regularly spaced whorls of lateral branches, each bearing a mass of spores. Photograph by Andy Swanson.

Plate 92. The tiny stalked fruiting bodies of *Protostelium mycophaga* are so small that they can be detected only with the use of a microscope. Photograph by John Shadwick.

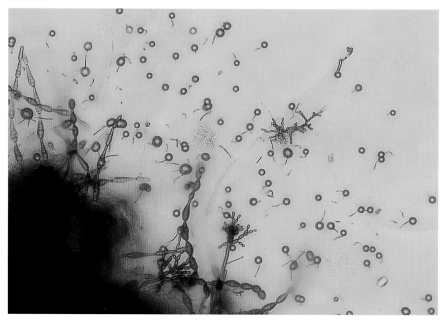

Plate 93. The fruiting bodies of *Cystoderma amianthinum* (saffron powder-cap) are commonly associated with bryophytes. Photograph by Emily Johnson.

Plate 94. Both the mantle and the hyphae that grow down among the cells of the cortex are visible in this cross section of an alder (*Alnus*) root. Photograph by Gary Laursen.

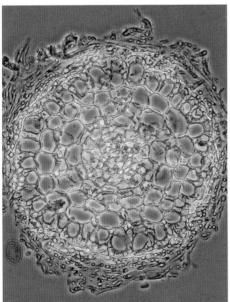

Plate 95. Hyphae of an endomycorrhizal fungi in the soil near the root of a host plant. Photograph by Joseph Morton.

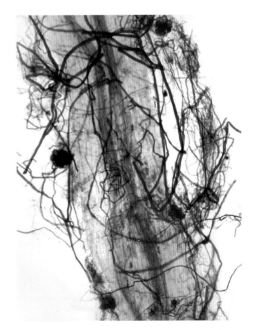

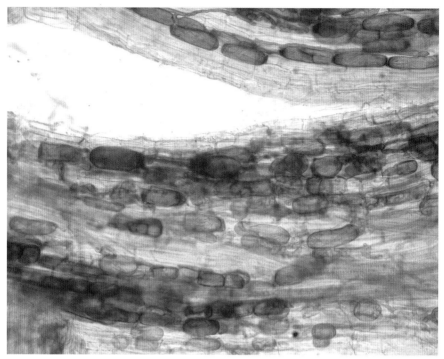

Plate 96. An example of a root in which numerous vesicles of an endomycorrhizal fungus are present. Photograph by Joseph Morton.

Plate 97. Broken spores of an endomycorrhizal fungus. Photograph by Joseph Morton.

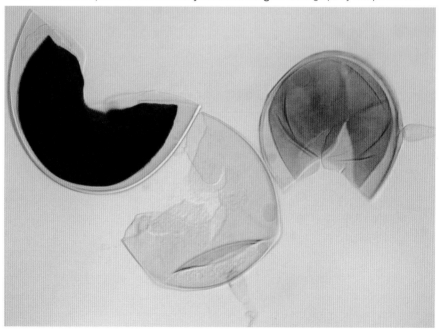

Plate 98. Because of its appearance, Indian pipe (*Monotropa unifora*) is a plant that is sometimes mistaken for a fungus.
Photograph by Charles Garratt.

Plate 99. *Puccinia podophylli* (mayapple rust) on the leaf of mayapple (*Podophyllum peltatum*).
Photograph by Emily Johnson.

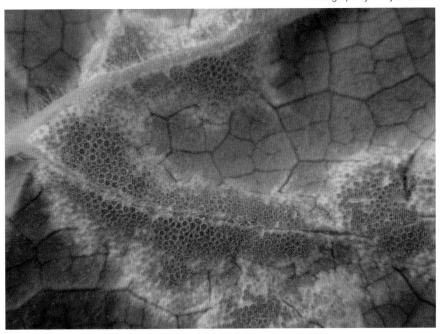

Plate 100. Wheat plant infected with *Puccinia graminis*. Photograph by Ronnie Coffman.

Plate 101. *Gymnosporangium juniperi-virginianae* (cedar apple rust) on red cedar (*Juniperus virginiana*). Photograph by Steven L. Stephenson.

Plate 102. A flower of white campion (*Silene alba*) infected by *Microbotryum dianthorum* (anther-smut) produces anther-like structures that contain fungal spores instead of pollen. Photograph by Michael Hood.

Plate 104. Gall-like swellings on a rhododendron that has been infected by a species of *Exobasidium*. Photograph by Emily Johnson.

Plate 103. An ear of corn infected by the corn smut fungus (*Ustilago maydis*). Although it may not look very attractive, this fungus is consumed by humans in some parts of the world. Photograph by Emily Johnson.

Plate 105. Two leaf-cutter ants carrying plant parts back to the nest. Photograph by Tom Pickering.

Plate 106. A large mound constructed by fungus-cultivating termites in the grasslands of the Republic of Namibia in Africa. Photograph by Scott Turner.

Plate 107. Fruiting bodies of the termite fungus (*Termitomyces*). The long stalks with their tapering rootlike extensions are clearly apparent. Photograph by Scott Turner.

Plate 108. Ambrosia beetle (*Xyleborinus saxesenii*) adults and larvae feeding on fungal hyphae and conidia. Photograph by Peter Bierdermann.

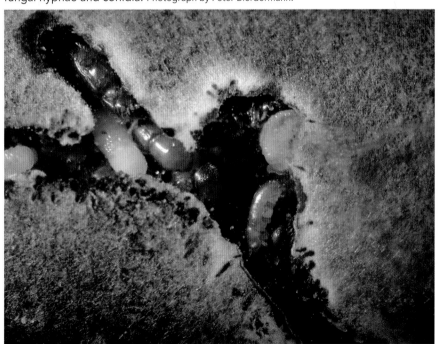

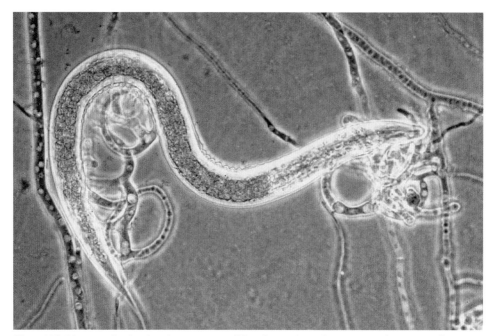

Plate 109. Nematode that has been trapped by the nematode-trapping fungus *Arthrobotrys oligospora*. This species uses an adhesive net to trap nematodes. Photograph by George Barron.

Plate 110. The long-footed potoroo is an obligate mycophagist in the forests of eastern Victoria and New South Wales in Australia. Photograph by Ben Wrigley, Houseman, and Wrigley, Department of Environment and Climate Change, New South Wales, Australia.

Plate 111. The bioluminescent fruiting bodies of *Mycena lucentipes* as they appear in total darkness. With a film camera and a long exposure time, this fungus can be photographed with its own light. Photograph by Cassius Stevani.

Plate 112. The world of fairies and the association of these mythical creatures with a ring of mushrooms (a so-called fairy ring) are depicted in the Victorian era painting entitled *A Fairy Ring*. Photograph supplied by Peter Nahum at the Leicester Galleries, London.

Plate 113. A "fairy ring" formed by the fruiting bodies of an agaric in a grassy lawn. Fairy rings are not uncommon in such settings. Photograph by Emily Johnson.

Plate 114. Colors produced from mushroom-derived dyes. Photograph by John Plischke.

Plate 115. *Pisolithus tinctorius* (dye-makers false puffball) can be used a a source of a rich golden-brown to black dye. This fungus often occurs in disturbed sites with poor soils. Photograph by Emily Johnson.

Plate 116. *Panaeolus foenisecii* (lawn mower's mushroom). The fruiting bodies of this fungus occur in small groups among the grasses in lawns through the Northern Hemisphere. Photograph by Emily Johnson.

Plate 117. *Amanita muscaria* (fly agaric), one of the best known of all macrofungi. Photograph by Emily Johnson.

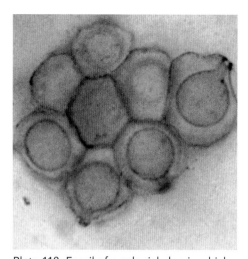

Plate 118. Fossil of a colonial alga in which each of the individual cells contains the zoosporangium of a chytrid. Photograph courtesy Forschungsstelle für Paläobotanik, WWU Münster and supplied by Hans Kerp.

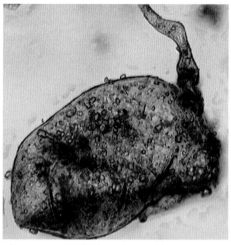

Plate 120. Fossil of a *Glomus*-like spore from the Late Ordovician. Photograph by Dirk Redecker.

Plate 119. *Palaeoblastocladis milleri* associated with the stem of *Aglaophyton major*. Two tufts of hyphae produced by this fossil chytrid are apparent in the cross section of the stem of this primitive plant. Photograph courtesy Forschungsstelle für Paläobotanik, WWU Münster and supplied by Hans Kerp.

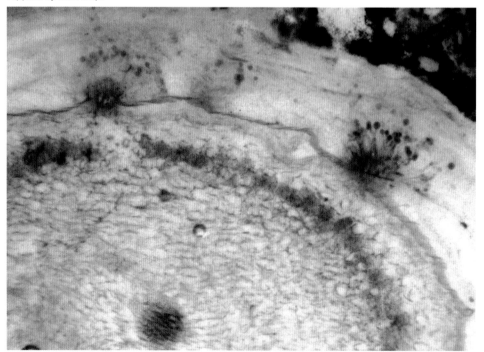

Plate 121. An exceptionally large specimen of *Prototaxites* discovered in northwestern Saudi Arabia. This specimen was 17 feet (5.3 m) long and 4 feet (1.37 m) in diameter at the base. Photograph by Charles Meissner.

Plate 122. Hypothetical reconstruction of an Early Devonian landscape. *Prototaxites* would have towered over the primitive vascular plants that existed at this time in the earth's history. Painting by Mary Parrish, Department of Paleobiology, Smithsonian Institution.

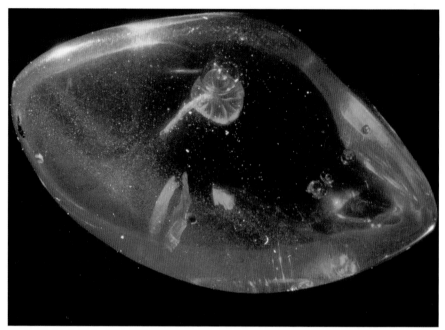

Plate 123. Fruiting body of *Protomycena electra* in a piece of amber from the Dominican Republic. Photograph by David Grimaldi.

Plate 124. A fruiting body of *Arcyria sulcata* in a piece of Baltic amber that is 35 to 50 years old. Photograph by Alexander Schmidt.

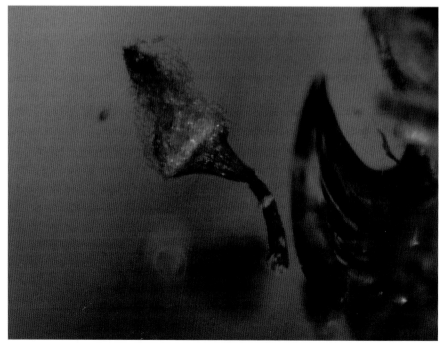

Distribution of Protostelids

Quantitative studies of the distribution and occurrence of protostelids in nature are a recent development, but the body of data accumulated thus far is extensive enough to allow some generalizations to be made. As a result of sampling carried out in a wide range of ecosystems, it is now known that protostelids can be found in virtually all terrestrial habitats. Some species, such as *Protostelium mycophaga*, are very common, whereas other species range from being uncommon to rare.

Most species appear to be cosmopolitan, since almost all of the described species have been recorded in all regions of the world that have been sampled intensively. Overall biodiversity of protostelids appears to decrease towards the poles, since samples collected in high latitude regions of the world tend to yield few protostelids.

Although the assemblages of species present in different habitats may be similar, differences often exist for different microhabitats in the same habitat. For example, certain species are present on bark or on dead aerial plant parts but tend not to occur in both microhabitats. Likewise, ground litter supports species that are uncommon on dead aerial plant parts. Interestingly, this distinction between ground and aerial microhabitats seems to be less pronounced in tropical forests than in temperate habitats (forests and grasslands).

Ecological Interactions Among Slime Molds

Since they apparently coexist in some of the same microhabitats and share the same primary food resource (that is, bacteria), the potential for ecological interactions among members of the three groups of slime molds would seem to exist. Unfortunately, there is very little information on this aspect of slime mold ecology.

In one study in which the vertical distribution of bacteria, myxomycetes, and dictyostelids was examined at three different depths in the soil/litter microhabitat of two deciduous forests in the mountains of southwestern Virginia, both myxomycetes and dictyostelids were present at all three depths, but their distribution patterns were not the same in the two forests. Numbers of either myxomycetes or dictyostelids were positively correlated with numbers of bacteria at every level in each forest, but the two groups never displayed a positive response at the same depth in the same forest. In other words, one group or the other displayed a response, but never both (Stephenson and Landolt 1996). Presumably, this would be the case if there was some degree of competition for the bacteria upon which both groups feed.

In another study in which selected species of myxomycetes and dictyostelids were co-cultured in all possible pairwise combinations in the laboratory, the myxomycete involved in some pairings exhibited delayed to occasionally nearly complete inhibition of plasmodial formation and development. Dictyostelids sometimes exhibit evidence of a killer activity between species, with the presence of one species having a negative effect on another (Hagiwara and Someya 1992). Perhaps the same factor may contribute to their apparent toxicity to some myxomycetes. In any case, there appears to be evidence for some degree of ecological interaction among different types of slime molds.

9

The Role of Fungi in Nature

Recyclers of Dead Organic Material 164
Decomposers of Wood 165
White-rot and Brown-rot Fungi 165
Soft-rot Fungi 167
Decomposers of Leaves and Nonwoody Plant Parts 167
Decomposers of Liverworts, Hornworts, and Mosses 168
Decomposers of Dung 169
Plant Symbionts 170
Ectomycorrhizae 171
Endomycorrhizae 172
Special Types of Mycorrhizae 174
Ericoid and Monotropoid Mycorrhizae 174
Orchid Mycorrhizae 176

Mycorrhizal Associations with Hornworts and Liverworts 176
Endophytic Fungi 177
Food for Animals 178
Plant Parasites and Pathogens 178
Rusts 178
Mayapple Rust 179
Wheat Rust 179
Cedar Apple Rust 182
Smuts 182
Anther-smuts of Flowers 183
Corn Smut 183
Ericoid Smuts 184
Loose Smut of Oats 184

There is little doubt that fungi play important but often underappreciated roles in nature. In the preceding chapters, many of these roles have been mentioned in the context of particular taxonomic groups of fungi.

Recyclers of Dead Organic Material

One recurring role has been fungi as saprotrophs. In both terrestrial and aquatic ecosystems, fungi are the organisms responsible for breaking down dead organic matter, particularly dead plant material. Dead organic material derived from animals is largely broken down by bacteria, although fungi can be involved in the break-down of some structural components of animals. These include collagen (an abundant structural protein in animals) and keratin (the substance of which hair is composed).

The degradation of dead plant material by fungi is absolutely essential for the continuation of life on the earth. The photosynthesis that takes place in living green plants and algae involves the fixation of atmospheric carbon dioxide into organic molecules. The green plants and algae then use some of these organic molecules as food to meet their own energy needs, but other organic molecules go into storage (as starch in the more familiar vascular plants) or are incorporated into new cells or tissue (growth or reproduction by the organism in question).

Animals (including humans) meet their energy needs by obtaining (usually by directly or indirectly consuming plants or plant parts) some of these organic molecules. As a general rule of thumb, perhaps 10 percent of the biologically fixed atmospheric carbon is passed along to animals, where it is released back into the atmosphere through respiration or (ultimately) the death of the animal in question. Something must happen for the other 90 percent of the biologically fixed atmospheric carbon contained in plants to be returned to the atmosphere, and this is where fungi become involved.

Fungi as a group have the capability to break down almost every type of organic substance (or substrate). A fungus accomplishes this by producing various digestive enzymes that are released into its immediate environment. Different fungi produce different enzymes and consequently, can degrade different substrates.

All fungi can produce the enzymes necessary to break down relatively simple organic molecules such as carbohydrates, and many can degrade and utilize cellulose, the most common organic substance on the earth, representing about a third of all plant matter. Far fewer fungi (mostly basidiomycetes along with a few ascomycetes) have the capability of decomposing lignin, which is only slightly less abundant than cellulose. Both cellulose and lignin are important components of wood. Typically, the wood of a tree consists of 40 to 60 percent cellulose and 15 to 30 percent lignin, with wood from conifers containing a relatively higher proportion of lignin than wood from woody angiosperms.

Decomposers of Wood

In a purely technical sense, the term "wood" refers to the xylem (water-transporting tissue) of a vascular plant that is produced as a result of secondary growth. Not all plants have secondary growth, and a distinction is often made between woody plants and non-woody (or herbaceous) plants in which only primary growth (essentially an increase in length of the main axis) takes place. Actually, all vascular plants start out as non-woody, when all of their cells have walls consisting predominantly of cellulose. As the plant grows and develops over time, the cell walls become thickened by the addition of more cellulose or, for those cells in parts of the plant specialized for support and the transport of water, lignin. Lignin is structurally more complex than cellulose and far more resistant to being decomposed by fungi.

For woody plants such as trees, new layers of xylem are added each year, gradually increasing the diameter of stems and roots. Secondary growth is most apparent in the stem, which is more commonly referred to as a trunk when the plant begins to reach some size. This process may continue for many years. In exceptional examples such as the bristlecone pine (*Pinus longaeva*), new layers of xylem can be added over a period of more than 4000 years.

For the most part, the cells that make up secondary xylem are dead, so the original nutrient-rich protoplasts (the cell contents) are no longer available as a resource that the hyphae of a fungus can exploit. As such, the fungi that meet their energy needs as saprotrophs of wood are a very different group from those found on living plants as saprotrophs or parasites. The latter will be considered in more detail later in this chapter. The enzymes required to degrade cellulose are different and (as already noted) much more universal than those required to degrade lignin.

White-rot and Brown-rot Fungi
Taxa of wood-decomposing fungi are often assigned to two categories on the basis of whether or not they can degrade both cellulose and lig-

nin or just cellulose alone. The members of the first category (the so-called white-rot fungi) have the enzymes necessary to degrade both cellulose and lignin more or less simultaneously. The residual material that is left behind has a somewhat fibrous appearance and is very pale in color, looking as if it had been bleached. In contrast, wood degraded by members of the second group (the so-called brown-rot fungi) is brown in color and tends to be broken up into somewhat cubical fragments that quickly disintegrate into a powdery brown residue. In both instances, the structural integrity of the wood is lost, which can pose a major problem if the piece of wood involved forms part of a building.

Serpula lacrymans (dry rot fungus) is considered the most economically important wood decay fungus in temperate regions of the world. Once this fungus gains a foothold in the wood of a building, it can spread very rapidly and cause a great deal of damage, often to the extent that much of the original wood has to be replaced. Like other similar wood-decay fungi, *S. lacrymans* cannot survive in wood with a level of moisture less than 20 percent, so the key is not to allow wood in any part of a building to remain damp for any period of time.

Other brown-rot fungi are common in nature, where they are associated with living and dead trees. Living trees can survive the presence of the fungus if the latter is confined to the wood in the center of the trunk. Over time, much of the central portion of the trunk can be lost to wood decay, leaving only a cylinder around what is essentially an open cavity. This is what accounts for a hollow tree. Often, the fact that a particular tree is hollow is not known until a strong wind applies enough stress to cause the weakened trunk to break off. It is not unusual for a significant proportion of the older trees in a forest to be hollow to at least some degree.

Among the more common examples of brown-rot fungi found in nature are *Laetiporus sulphureus* (sulfur shelf, PLATE 57) and *Piptoporus betulinus* (birch polypore). Fruiting bodies of the latter occur on dead and dying birch trees throughout the Northern Hemisphere. There are more species of white-rot fungi than there are of brown-rot fungi. Widespread and common examples of white-rot-fungi mentioned earlier in this book are *Daldinia concentrica* (carbon balls, PLATE 45) and *Trametes versicolor* (turkey tail, PLATE 58). Brown-rot fungi are especially common on conifers and are most abundant in cool temperate regions of the Northern Hemisphere, while white-rot fungi can be found throughout the world.

Soft-rot Fungi

The term "soft-rot" is sometimes applied to situations in which only the outermost layers of wood are subject to decay. Soft rots occur in wood that has an unusually high level of moisture. Wood inside buildings that have been subjected to floods or are water-logged for some other reason is particularly susceptible. Most of the fungi involved are ascomycetes, with species of *Chaetomium* among the most common and best known. Only the cellulose of the wood is degraded by soft-rot fungi. Interestingly, the plasmodia of some myxomycetes (for example, *Physarum polycephalum*) have been demonstrated to produce the enzymes (called cellulases) necessary to break down cellulose. Their ecological impact as decomposers in natural ecosystems is probably insignificant, but myxomycetes do occur on many of the same substrates as wood-decay fungi.

Decomposers of Leaves and Nonwoody Plant Parts

Leaves fruits, and seeds represent a different type of substrate than wood or bark. Considerable diversity exists for the leaves produced by plants, and what might be termed a typical leaf from a common and widespread broadleaf tree is markedly different from the very large leaves of palms and tree ferns or the needlelike leaf characteristic of many conifers.

The spores of fungi can reach a leaf while it is still on the tree, and the extent to which the leaf serves as a "spore trap" is related to such factors as its size, position on the tree, surface (smooth or hairy), and whether or not the spore lands when the leaf is wet. Studies have shown that numerous spores are already present on older leaves prior to leaf-fall. These are carried with the leaf when it falls to the ground.

The time required for a particular leaf to decompose completely varies from habitat to habitat and for different types of leaves. In temperate forests, the leaves of most broadleaf trees last for less than a year, while the needlelike leaves of conifers usually decompose over several years. In a moist tropical forest, most leaves do not persist for more than a few weeks, but those of palms and tree ferns represent an exception and can last much longer.

Once on the ground, a leaf can be invaded by soil-inhabiting fungi or by those already present in the layer of litter at the soil surface. Careful examination of a handful of leaf litter will invariably reveal the presence of fungal hyphae. As was noted in Chapter 6, the fruiting bodies of certain agarics (for example, species of *Marasmius* and *Collybia*) often occur in some abundance on dead leaves, which simply reflects their biological role as litter-inhabiting saprotrophs.

The fungi associated with fruits and seeds are not necessarily the same ones found on leaves. In some cases, a particular species may be restricted to the substrate represented by a certain type of fruit or seed. Two examples are *Mycena luteopallens* (walnut mycena), which occurs only on the husks of hickory and walnut, and *Strobilurus conigenoides* (magnolia-cone mushroom), a species found only on old fruits of magnolia. Another species apparently confined to a specific substrate is *Baeospora myosura* (conifer-cone baeospora), which is commonly encountered on the cones of various conifers.

Decomposers of Liverworts, Hornworts, and Mosses

Although they receive much less attention than the fungi that decompose wood or litter, there is a group of fungi which decompose bryophytes. In some habitats where bryophytes are abundant, such as in many high-latitude regions of the world (Horak and Miller 1992), it would be a mistake to dismiss the ecological importance of these bryophilous ("moss-loving") fungi. On subantarctic Macquarie Island, where Gary Laursen, a mycologist at the University of Alaska, and I carried out a survey for fungi in 1995, the fruiting bodies of a number of species of *Galerina* were commonly collected from the extensive mats of bryophytes that occur at lower elevations on the island.

Arrhenia retiruga (sometimes listed as *Leptoglossum retirugum*) is a relatively common bryophilous fungus in the Northern Hemisphere. The gelatinous cap of *A. retiruga* is usually no more than about ³⁄₈ inch (1 cm) in diameter, and the stalk is short or lacking. As such, the fruiting bodies are small and not easy to spot in the field. This is not the case for *Cystoderma amianthinum* (saffron powder-cap), the fruiting bodies of which can be 3 inches (8 cm) tall and have a cap 1⅝ inch (4 cm) in diameter (PLATE 93). This fungus is common and widespread in temperate and boreal regions of the world, where it often occurs in small clusters on mats of bryophytes associated with the forest floor. As a group, the bryophilous fungi have received relatively little study, and it is likely that there are still numerous species as yet unknown to science.

Decomposers of Dung

The fungi associated with the dung of herbivores represent an ecologically important but underappreciated group. The substrate involved, if considered objectively, is undigested plant material to which animal waste products have been added. The exact nature of dung depends on the efficiency of the digestive tract of the herbivore from which it was derived. Some animals such as cattle produce fine-textured dung consisting largely of fibrous plant material, while horses, which have a somewhat less efficient digestive system, produce much coarser dung. Moreover, the original type of plant material consumed by the herbivore is reflected in the composition of dung.

Once deposited, dung decomposes rapidly because the material of which it is composed is already broken up, has a high nitrogen content (as much as four percent), and is usually very moist. As such, dung is a particularly rich, although spatially limited and relatively ephemeral, substrate for fungi and other organisms.

The fungi that utilize dung are taxonomically diverse and include examples from most of the major phyla. The Zygomycota, Ascomycota, and Basidiomycota are well represented among the coprophilous ("dung-loving") fungi. For example, approximately 175 genera of ascomycetes include species that are found largely or exclusively on dung, and many of the species in what has been recognized as the genus *Coprinus* in the basidiomycetes are coprophilous.

As described earlier in this book, some coprophilous fungi (for example, *Pilobolus* in the Zygomycota) are highly specialized for survival in their unique habitat. These fungi have evolved ways of projecting their spores some distance beyond the dung upon which they occur, greatly increasing the likelihood of the spores being ingested by another herbivore. In many instances, the spore-producing structure of the fungus in question exhibits a positively phototropic response that determines the direction the spores will be projected. Moreover, the spores of many coprophilous fungi will not germinate until they have passed through the digestive tract of a herbivore.

Mycologists have often noted that the different taxonomic groups of coprophilous fungi display a fairly definite ecological sequence. Members of the Zygomycota (for example, *Pilobolus*) are the first to appear, usually within a matter of a couple of days. These are followed by members of the Ascomycota (for example, *Ascobolus*), which eventually are replaced by members of the Basidiomycota (for example, *Coprinus*). This sequence is not absolute, but it is so predictable that a common laboratory exercise in mycology involves having students monitor a culture of dung over a period of several weeks to observe the succession of fungi. Students are usually amazed at the number and variety of fungi they can observe in a single culture.

Many coprophilous fungi are generalists and are likely to occur on any type of herbivore dung, but a few seem to display a preference for the dung of particular kinds of animals (Kendrick 2000). Although limited in scope to just one phylum, the book *An Illustrated Guide to the Coprophilous Ascomycetes of Australia* by Ann Bell (2005) is a beautifully illustrated treatment of some of the fungi that occur on dung.

Plant Symbionts

A second major role, restricted to certain groups of fungi, involves the formation of intimate associations with plants. These associations, called mycorrhizal associations or mycorrhizae (literally meaning "fungus root"), are exceedingly widespread and are known to occur in 80 to 90 percent of all vascular plants, In brief the fungus benefits the plant by increasing the plant's ability to take up water and nutrients such as nitrogen, phosphorus, potassium, and calcium from the soil, while the plant benefits the fungus by supplying it with organic molecules produced as a result of photosynthesis.

Studies have shown that there are several other aspects to the association, especially with respect to the advantages it confers to the plant. For example, the presence of the fungus apparently increases the plant's tolerance to environmental stresses such as drought and reduces the chances of the plant being infected by other potentially harmful fungi.

Traditionally, mycorrhizal associations formed between plants and fungi have been divided into two types on the basis of structural differences in the way the association is formed. In the first type, referred to as ectomycorrhizae, the bulk of the fungus involved is on the outside the root and its presence is often fairly conspicuous. In the second type, referred to as endomycorrhizae, the bulk of the fungus is inside the root.

Ectomycorrhizae

Of the two types of mycorrhizal associations, endomycorrhizae are by far the more common. Only about three percent of plants form ectomycorrhizae, but some of the plants involved are rather common. In fact, ectomycorrhiza-forming trees are predominant in temperate and boreal forests of the Northern Hemisphere and often represent important components of forests in temperate regions of the Southern Hemisphere. Among the more common trees in these forests are oak, beech, birch, pine, spruce, and fir in the Northern Hemisphere and southern beech and eucalypts in the Southern Hemisphere.

Most of the fungi that form ectomycorrhizae are basidiomycetes, but there are also a few ascomycetes involved. Among the more commonly encountered ectomycorrhiza-forming basidiomycetes are species of *Amanita*, *Boletus*, *Cortinarius*, *Lactarius*, and *Russula*. Some species are relatively host specific and will form associations only with certain plants, but others appear to be generalists and have been recorded for many different plants.

An ectomycorrhizal association begins when the hyphae of an ectomycorrhiza-forming fungus come into contact with the root system of a potential host plant. These hyphae appear to be attracted to actively growing root tips, possibly in response to the presence of substances being exuded by the root tips. Once in contact with the root, the hyphae proliferate to give rise to a covering (called a mantle or sheath) that extends over the outside of the root (PLATE 94). The thickness of the covering varies for different ectomycorrhiza-forming fungi but often comprises 20 to 30 percent of the total volume of those portions of the root where the mantle is fully developed.

In many examples, the mantle is differentiated into two layers. The outermost layer consists of relatively thick-walled hyphae that are tightly packed together with few free spaces. In contrast, the innermost layer is composed of relatively thin-walled hyphae that are loosely packed together. The innermost layer of the mantle gives rise to hyphae that penetrate the root, growing down among the cells in the outermost tissues (the cortex) of the root to create what is referred to as the Hartig net. Other hyphae, produced by the outermost layer of the mantle, extend out into the soil.

The fungal mantle of an ectomycorrhizal root can be relatively conspicuous, but what it actually represented was not known until about

1885. Theodor Hartig, a German forestry scientist, was the first to describe and illustrate both the mantle and the hyphal network that forms inside the root in a paper published in 1840. The hyphal network was later named the Hartig net in his honor. However, he failed to recognize the fungal nature of these structures, attributing their origin to the root itself (Hartig 1840).

More than 40 years later, Albert Frank, a German botanist, pointed out that what we now know as ectomycorrhizae were widespread on woody plants in nature, and he went on to hypothesize that they represented a mutually beneficial association between the root system of the plant and fungi. The information presented in his paper (Frank 1885), which did not conform to the conventional wisdom of the time, was initially opposed by much of the scientific community. Indeed, it provoked a controversy that lasted well into the early twentieth century. Although ultimately accepted in the face of overwhelming observational and experimental evidence, the concept of mycorrhizae (including both ecto- and endomycorrhizae) was slow to find its way into textbooks of biology and ecology, and even today it often receives far less attention than is warranted as a result of its extraordinary importance to terrestrial ecosystems (Trappe 2005).

Endomycorrhizae

Endomycorrhizal associations do not involve the formation of a sheath around the root of the host plant. In fact, they are not usually possible to detect with the naked eye. The only indication of an endomycorrhiza-forming fungus tends to be the presence of a loose and very sparse network of usually rather thick-walled and nonseptate hyphae in the soil near the root, although it is often possible to identify one to several thick-walled, brown to black, sporelike structures connected to some of the hyphae (PLATE 95).

When the hyphae of an endomycorrhiza-forming fungus first contact the root of a potential host plant, they penetrate the outmost layer of cells (the epidermis) and then form numerous branches that grow between and also into individual cells of the cortex. Eventually, hyphae of the fungus may proliferate throughout the entire cortex, but they do not disrupt the vascular tissue at the center of the root.

Once inside an individual cell, particular hyphae form very finely branched, treelike structures called arbuscules. These often fill the interior of the cell. The branches of the arbuscules represent the primary sites of the exchanges of water, nutrients, and organic molecules that take place between the fungus and the host. The arbuscules are not permanent structures and usually last only a few days, after which the cell returns to normal.

In many endomycorrhiza-forming fungi, a second type of structure, called a vesicle, is produced within the root. Vesicles develop at the tips of individual hyphae. They are thin-walled, ovoid or spherical, and appear to serve a storage function for the fungus. In some host plants, they can be exceedingly numerous (PLATE 96). Because they involve the production of arbuscules and vesicles, endomycorrhizal associations of the type described above are often referred to as vesicular-arbuscular mycorrhizae (or VAM). Since some fungi produce arbuscules but not vesicles, this term is not entirely appropriate.

While some hyphae of the endomycorrhiza-forming fungus are proliferating inside the root, other hyphae are produced that extend out into the soil. These hyphae function in the same manner as described for ectomycorrhiza-forming fungi by greatly increasing the volume of soil from which water and nutrients can become available to the host plant. In addition, these hyphae also produce sporelike structures. The latter are relatively large (usually 0.016 to 0.24 inch [40 to 600 µm] in diameter) and contain hundreds of nuclei and large amounts of lipids as a food reserve.

Although certainly different from any of the spores described elsewhere in this book, the sporelike structures found in endomycorrhiza-forming fungi are probably best regarded as representing a special type of asexual spore. In the soil, spores may occur singly or in aggregations (called sporocarps) that sometimes reach a diameter of ¾ inch (2 cm). An individual spore often resembles a tiny balloon, with the terminal portion of the hypha from which it was derived still attached (PLATE 97). Spores range in color from hyaline to black, and the spore wall may be ornamented in some fashion. They are able to persist in the soil for a number of years.

The taxonomy of VAM fungi is based almost entirely upon the morphology of their spores. Because they are so large, spores can be separated from soil by a flotation technique that involves adding a small sample of soil to a container of water and then stirring the mixture thoroughly to separate the particles of soil and dislodge the spores.

Once they are free of the soil particles that surround them, many of the spores will float the top of the water, where they can be collected.

Special Types of Mycorrhizae

Although the vast majority of mycorrhizal associations conform to the general descriptions of the two types described thus far, there are actually some special types that are distinctly different. These are sometimes considered as unusual variants of endomycorrhizae, but one type (known as ericoid mycorrhizae) actually shares some features in common with ectomycorrhizae. Indeed, ericoid mycorrhizae were once referred to as ectendomycorrhizae because of this fact.

Ericoid and Monotropoid Mycorrhizae

Ericoid mycorrhizae receive their name from the fact that they occur only in a single family of vascular plants, the Ericaceae (heaths). Members of the Ericaceae have root systems in which the smaller roots are very fine and hairlike, with a cortex composed of one or a few layers of cells. Roots infected by the mycorrhizal fungus have both a sheath and a cortex in which the individual cells have densely coiled fungal hyphae present. In some instances, as much as 80 percent of the volume of the root is composed of fungal hyphae. The fungi involved in ericoid mycorrhizae are ascomycetes, with *Hymenoscyphus ericae* the best-known example.

Some ericoid mycorrhizae are distinct enough to be recognized by some mycologists as a different type. These are the monotropoid mycorrhizae found in members of the family Monotropaceae, once thought to be part of the Ericaceae but now considered to constitute a separate family.

Throughout temperate regions of North America and eastern Asia, one member of the Monotropaceae is sometimes mistaken for a fungus (PLATE 98). This plant (*Monotropa uniflora*) has been called by various names, including ghost flower, ghost pipe, corpse plant, or more commonly Indian pipe. It has a white, waxy appearance, usually occurs in the deep shade of the forest interior, is about the same size (6 to 8 inches [15 to 20 cm] tall) of many forest fungi, and grows up through dead leaves on the forest floor just like the fruiting body of a fungus. In fact, it sometimes appears on display tables at fungal forays. However, Indian pipe is not a fungus. Instead, it is a very unusual vascular plant.

The epithet "uniflora" means single flower, and each stem of the plant bears a solitary flower.

When the plant first emerges from the litter layer, the top of the stem is curved downward so that the flower faces the ground. The resemblance of the plant to a small, slender pipe accounts for its most widely used common name. Later, after pollination takes place, the stem grows in such a way that the flower turns upward. At this time, the plant turns pale pink to sometimes deep red. Eventually, after the stems die, they become black.

Indian pipe may occur as a single stem (which causes the plant to resemble a fungus even more) or in clumps of several stems. Scattered, scalelike leaves occur along each stem. In the soil, the stem arises from a mass of coralloid roots termed a root-ball. The latter is covered with a thick mantle composed of fungal hyphae, and other hyphae extend into the outermost layers of the cortex of the root. These hyphae penetrate some of the cells, where they form simple peglike structures and not the dense coils more typical of ericoid mycorrhizae.

Indian pipe is unusual because it lacks the green pigment chlorophyll that is necessary for photosynthesis. Although not the only vascular plant without chlorophyll, the pure white color of Indian pipe certainly underscores the "non-green" condition.

When I first learned about Indian pipe more than 40 years ago, the plant was described as a saprotroph, living on dead leaves and other plant material on the forest floor. However, this is not the case at all. Indian pipe is actually a mycoheterotroph. What this means is that the plant depends upon a nearby tree to meet its energy needs. This energy is transferred via a shared mycelial network that involves the tree, a fungus that has formed a mycorrhizal association with the tree in question, and the roots of the Indian pipe plant. The fungus obtains organic molecules produced as a result of photosynthesis from the tree, which is the normal situation in a mycorrhizal association. The fungus then transfers some of these organic molecules through its mycelium to the roots of the Indian pipe. There is no direct connection of the Indian pipe to the tree. Instead, the two are linked through the fungus.

It is difficult to see just what the fungus gets out of this association, but Indian pipe is not the only plant in which it is known to occur. For Indian pipe, most of the mycorrhizal fungi that link the plant to nearby trees appear to be species of *Russula* or *Lactarius* (Yang and Pfister 2006).

Orchid Mycorrhizae

The second special type of mycorrhizal association is exhibited by members of the vascular plant family Orchidaceae, the plants we know as orchids. This is one of the largest of all plant families, and more than 20,000 species have been described, the majority of which are tropical.

Orchid mycorrhizae are morphologically similar to ericoid mycorrhizae in that hyphae of the fungus involved form intracellular coils within individual cells of the host plant. Each intracellular coil has a short life, lasting only a few days before it degenerates and is absorbed by the cell. However, fungal hyphae occur in a considerable number of cells at any single point in time, and the mycelium from which they are derived extends well out into the soil.

Orchid mycorrhizae differ from other types in one fundamental way—the fungus supplies the plant with organic molecules instead of the other way around. These organic molecules are obtained by the fungus from the breakdown of organic debris in the soil or from linkages established with the root systems of nearby plants. The orchid uses the organic molecules provided by the fungus to meet its own energy needs, so the fungus is "feeding" the plant.

All orchids are dependent upon their fungal associate during the early stages of their life cycle, and for a few species of orchids, this dependence continues throughout the entire life of the plant. The best examples of such orchids are those species that lack chlorophyll. *Corallorhiza maculata* (spotted coral root), a species that is occasionally encountered in the forests of eastern North America, represents one example. This orchid characteristically occurs in areas of moist forest where there is a thick layer of dead leaves and humus. In late summer, the orchid sends up a slender purple to bronze stem that can be as much as 1 to 2 feet (30 to 60 cm) tall. Arising from the top of the stem is an elongated cluster of small flowers that are about the same color as the stem except for a lower petal, which is white with purple spots (which accounts for the common name of the plant). As is the case for *Monotropa uniflora*, *C. maculata* is a mycoheterotroph.

Mycorrhizal Associations with Hornworts and Liverworts

Mycorrhizal associations are not limited to vascular plants but also occur in two groups of bryophytes, the hornworts and liverworts (Read et al. 2000). This fact is often overlooked when the subject of mycorrhizae is considered. For example, species of *Hymenoscyphus*, so prominent in ericoid mycorrhizae, have been reported as associates of leafy liverworts in the families Adelanthaceae, Cephaloziellaceae, Cephaloziaceae, Calypogeiaceae, and Lepidoziaceae. The fungi produce structures in the rhizoids of the liverworts that appear similar to those found in the Ericaceae. The species of liverworts forming these "ericoid" mycorrhizae often form mats in sphagnum bogs where conifers such as spruce and shrubs belonging to the Ericaceae also occur. As such, it is possible that complex linkages involving mycorrhizal fungi exist among the liverworts, ericaceous plants, and conifers.

Other liverworts in the families Lophoziaceae, Arnelliaceae, and Scapaniaceae also form mycorrhiza-like associations with fungi. The taxa involved belong to the Basidiomycota, and some are known to be ectomycorrhizal. Although mycorrhizae appear to be common and widespread in liverworts, they have received relatively little study. It is perhaps ironic, as will be mentioned in Chapter 12, that bryophyte-like plants (which predated vascular plants on land) may have been involved in the very first plant-fungus associations. In this context, it is interesting to note that the liverworts are considered the most primitive of the bryophytes. Mycorrhiza-like associations have yet to be documented for mosses, the most advanced group of bryophytes.

Endophytic Fungi

Mycorrhizae are not the only intimate associations formed by fungi and higher plants. Some microscopic ascomycetes, a group of fungi known as endophytes, actually live inside the stem and leaves of plants. The exact nature of the relationship between endophytic fungi and their plant hosts is not yet completely understood, but the fungi generally do not appear to harm their hosts. Indeed, there is some evidence that the presence of endophytic fungi can cause the plant to have a greater resistance to insects, roundworms, and bacteria. Moreover, these fungi seem to be able to promote the production of poisonous alkaloids, chemicals that can deter the consumption of plant parts by herbivorous (plant-eating) mammals.

Food for Animals

A third role that fungi play is essentially a consequence of their first two roles, since the mycelia and fruiting bodies of saprotrophic and mycorrhizal fungi provide a source of food for a very diverse assemblage of animals, ranging from insects and other invertebrates such as slugs and snails to rodents and even large mammals. The "who eats whom" equation also works the other way, since there are (as described in Chapter 10) "carnivorous" fungi that have developed hyphal traps in which they can capture small protists such as amoebae, as well as nematodes, rotifers, and tardigrades.

Plant Parasites and Pathogens

Fungi also have an important role as plant parasites and pathogens. Fungi rather than bacteria are the most widespread and destructive parasites of plants whereas the reverse is true for animals.

As plant parasites, fungi meet their energy needs by deriving organic materials from their hosts, to the detriment of the latter. In many instances, the host plant appears to tolerate the parasitic fungus with little evidence of any harmful effect. In other instances, the presence of the fungus is all too apparent, and the plant is then considered diseased. Clearly, the host plant is being harmed. If this condition persists and the plant becomes so weak that it subject to dying, then the fungus has crossed the threshold that separates parasite from pathogen.

The distinction between the two is not always clearly drawn, since a particular fungus can be a rather innocuous parasite under some circumstances and a virulent pathogen on others. In fact, some fungi that are normally saprotrophs can convert to being pathogenic. For example, the dead wood of an otherwise still-living tree can be invaded by various wood-decomposing polypores, which may live on the trees for years.

Rusts

Fungi belonging to the subphylum Pucciniomycotina (rusts) in the Basidiomycota do not produce macroscopic basidiocarps, and all are obligate parasites of vascular plants. They usually do not kill their hosts (which would mean that the fungus would die, too), but rusts considerably reduce the yield of the plants they infect, and some of the latter are agriculturally important. Among these are wheat and other cereal grains.

The vast majority of rusts have very narrow host ranges, being restricted to a single family, a single genus, or even a single species. Because of this, early mycologists often routinely described new species of rusts whenever they encountered this type of fungus on a plant not previously reported in the literature as a host. Potential hosts for rusts include a variety of different plants, ranging from ferns to gymnosperms and angiosperms. As a group, these fungi are characterized by exceedingly complex life cycles, and the usual situation is for two different (sometimes distantly related) taxonomic groups to be involved. For example, the very common cedar apple rust, to be discussed below, alternates between a gymnosperm and an angiosperm.

The other distinctive feature of their life cycle is that as many as five different kinds of spores are produced, with each type specialized for a particular stage. One type is the basidiospore, which in rusts is actually derived directly from another type of spore (called a teliospore) that overwinters in ground litter.

Mayapple Rust

Puccinia is the largest genus of rust fungi, and some species are exceedingly common. Most of these are likely to be overlooked in nature and have little or no importance in human affairs. One example is *P. podophylli* (mayapple rust), which infects the leaves of *Podophyllum peltatum* (mayapple). The host plant is often abundant in low, moist areas throughout much of eastern North America. Leaves of infected plants clearly reveal the presence of the fungus (PLATE 99).

Wheat Rust

While *Puccinia podophylli* is of little consequence to humans, this is certainly not the case for *P. graminis* (wheat rust). As is true for most rusts, the life cycle of *P. graminis* is complex and involves two host plants, in this case wheat and either barberry (*Berberis vulgaris*) or one of its close relatives. It is the presence of the fungus on wheat that is of far greater concern. When *P. graminis* infects a wheat plant, the mycelium (which is dikaryotic at this point) of the fungus spreads throughout the tissues of the host.

The first indication of the infection is the appearance of numerous pustules (uredinia) on the stem and leaves of the wheat. These uredinia (singular: uredinium) represent places where masses of what are known as urediospores are formed. The urediospores are brick-red in color, and a heavily infected field of wheat appears rust colored, which suggests how *Puccinia graminis* and other members of the group to which it belongs received their common name (PLATE 100).

Urediospores are dispersed from the uredinia by wind, which can carry them to new hosts. One uredinium can produce many thousands of urediospores, and there may be several generations of urediospores during a growing season. In regions of the world where winters are mild, urediospores can survive and infect new host plants the next growing season. However, the next stage in the life of *Puccinia graminis* involves the production of another type of spore, the teliospore.

Late in the year, the uredinia begin to produce two-celled teliospores instead of urediospores, converting into what are known as telia (singular: telium). Teliospores remain attached to the host plant and are commonly left in the field on the straw residue from the wheat plants. Each of the two cells of a teliospore is initially dikaryotic, but the two nuclei eventually fuse to produce a single diploid nucleus. Teliospores represent the most resistant stage in the life cycle and remain dormant throughout the winter months. In the spring, the teliospores germinate and each of the two cells immediately gives rise to a special type of four-celled basidium that has basidiospores. During this process, the diploid nucleus in each cell undergoes meiosis, and each of the four resulting haploid nuclei is incorporated into a developing basidiospore.

Two different mating types exist for *Puccinia graminis*, and each is represented by half of the basidiospores. Mature basidiospores are forcibly ejected from the basidium and carried away by air currents. They are short-lived and usually do not travel very far. Basidiospores are unable to infect wheat plants and require barberry to serve as an alternate host. If there are no barberry plants available, the life cycle cannot continue. However, if a basidiospore does reach and then infect barberry, the mycelium of the fungus proliferates throughout the leaves of the host plant.

A basidiospore is haploid, but the dikaryotic condition is restored when an exchange of nuclei occurs between two compatible mycelia, with each derived from one of the two different mating types. This means that more than a single basidiospore must infect the same leaf of the barberry plant in order for the life cycle to continue. The exchange of nuclei typically involves the production of flask-shaped reproductive structures called spermogonia (singular: spermogonium), but simple fusion of vegetative hyphae from the two different mycelia may occur on some occasions. Eventually, specialized hyphae from the newly produced dikaryotic mycelium produce chains of spores in structures called aecia (singular: aecium). These spores, called aeciospores, are liberated and carried away on air currents. Upon making contact with a suitable host, an aeciospore germinates and gives rise to a mycelium, thus initiating the primary infection of the wheat plant.

Because of the economic importance of wheat, efforts to eradicate the alternate host would seem to represent a practical method to control the wheat rust disease. The connection between wheat and barberry had long been suspected, and the first laws relating to the eradication of barberry were implemented well before the German mycologist Heinrich Anton de Bary actually documented, in the mid-nineteenth century, that different stages of the same fungus were present on the two plants. A major eradication effort in the United States began in 1918 and lasted for several decades. Overall, this effort was judged to be quite successful, since the severity of the wheat rust disease was considerably reduced.

The rusts that infect wheat and other cereal grains have been intensively studied. Although the fact that they were fungi was not understood, their impact upon crops almost certainly has been recognized since the advent of agriculture thousands of years ago. The ancient Romans were very much aware of plant diseases caused by rusts and considered one of their lesser gods to be responsible. Since all of the major cereal grains (wheat, rice, oats, barley, and corn) that supply a major proportion of the food needed to support humans on the earth are hosts for rusts, one might speculate as to what the potential consequences would be if all cereal grains were to fall victim to these fungi.

When I was still a student in high school, I read a science fiction novel with a story line that described this type of hypothetical situation, albeit with a virus and not a fungus as the virulent pathogen involved. The novel, published in 1965, was entitled *The Death of Grass* (subsequently published in the United States as *No Blade of Grass*) and was written by the British author Samuel Youd (under the pseudonym of John Christopher). The story line is developed around the concept of the sudden appearance of a virus that kills all grasses.

As the story opens, the virus has already attacked rice in East Asia, causing massive famines. Next to be attacked are wheat and barley in places such as Europe. The total devastation of grasses by the virus seemed likely to cause a massive worldwide famine. The story follows the trials and struggles of the narrator's family as they attempt to make their way across the United Kingdom, which is already descending into anarchy, to the safety of his brother's potato farm. The potato, which is not a grass, would not be affected by the virus. I have always been a fan of science fiction, and I found this novel to be especially intriguing. Although totally fictional, the concept developed in the novel does provide some indication of the level of dependence (either directly or indirectly) of humans upon cereal grains.

Cedar Apple Rust

One of the more conspicuous rusts in many parts of North America is *Gymnosporangium juniperi-virginianae* (cedar apple rust) which occurs on red cedar (*Juniperus virginiana*) throughout the range of the latter in North America (PLATE 101). Red cedar is the primary host for this rust, and the alternate host is either apple or one of its close relatives. The presence of the fungus on the primary host results in the formation of a number of galls near the tips of smaller branches on the tree. Each gall consists of a combination of host tissue and the mycelium of the fungus and can be as much as an inch (2.5 cm) or more in diameter. In the spring, numerous hornlike or fingerlike masses of reddish orange teliospores (referred to as telial horns) began to protrude from the surface of the gall. These telial horns eventually become gelatinous and can reach a length of as much as an inch (2.5 cm). When the telial horns are most apparent, the infected cedar tree has the general appearance of a Christmas tree with decorations.

Smuts

Members of the subphylum Ustilaginomycotina, the fungi commonly known as smuts, represent a much smaller group than the Pucciniomycotina. These fungi are similar to the rusts in that they do not produce fruiting bodies and are all obligate parasites of plants, although only a single host plant is involved. In spite of their similarities, smuts and rusts are not closely related.

Smuts derive their name from the dark, dusty or sooty masses of teliospores that form on diseased plant parts. Although the mycelium of some species can spread throughout an infected plant, those of other species are confined largely to specific parts. Often, only the flowers show clear evidence of the presence of the fungus. This is the case for the disease caused by the anther-smut *Microbotryum dianthorum*.

Anther-smuts of Flowers

The fungus *Microbotryum dianthorum* (sometimes listed as *Ustilago violacea*) infects species in the plant family Caryophyllaceae (pinks), one of which is *Silene alba*, commonly called white campion (Thrall et al. 1993). In this plant, the presence of the mycelium of *M. dianthorum* causes the flowers to produce antherlike structures that contain teliospores instead of pollen (PLATE 102). The fungus also prevents normal development of the ovary of the flower. Insect pollinators that visit the diseased flowers pick up some of these teliospores. The insect then deposits the teliospores on the flowers of healthy plants of the same species, causing these plants to become infected with *M. dianthorum*. Ultimately, an entire population of *S. alba* can be affected (Alexander et al. 1996).

Silene alba and *Microbotryum dianthorum* represent an example of a host-pathogen system that was examined rather extensively during the late 1980s and throughout the 1990s at the University of Virginia Mountain Lake Biological Station in southwestern Virginia. I spent a number of summers at the station during this period and thus had an opportunity to observe first-hand the studies being carried out by researchers and students making what was referred to at the time as the "silene group."

Corn Smut

Teliospores are the conspicuous spores produced by smuts, and these "smut spores" as they are called, are exceedingly important for identification. Teliospores are thick-walled and often darkly pigmented. Many have a spore wall with a distinctive ornamentation, and some occur in clusters called spore balls. Teliospores represent the overwintering stage in the life cycle of smuts, and those of some species are known to be capable of surviving several years in the soil.

Among the best-known examples of this is corn smut, caused by the fungus *Ustilago maydis*. Corn smut is found throughout the world wherever corn is grown. In infected plants, the disease causes some or all of the kernels on an ear of corn to be replaced by grossly swollen masses of black teliospores (PLATE 103).

Although the presence of this disease causes significant losses in corn as a crop, this is not always an undesirable thing. Surprisingly, before they become fully mature, infected ears of corn can be collected and the galls (known as *cuitlacoche* or "maize mushrooms") sold as food in some parts of the world. In Mexico, they are considered a delicacy. Once, on a collecting trip to Mexico, I had the opportunity to sample *cuitlacoche* and found it to be somewhat grainy in texture and relatively tasteless. In any case, it certainly is not the first thing that comes to mind when most people think about edible fungi.

Ericoid Smuts

Some of the more conspicuous and distinctive looking members of the subphylum Ustilaginomycotina are found in *Exobasidium*, a genus of about 50 species. These fungi occur as parasites of plants in the family Ericaceae, which includes blueberry (*Vaccinium*) and azalea (*Rhododendron*). Infected parts (usually leaves) of the host plant develop conspicuous gall-like swellings (PLATE 104). These structures are commonly encountered on both blueberry and azalea in the upland forests of the Appalachian Mountains in eastern North America, where the host plants are often abundant.

Loose Smut of Oats

Another widely known smut is *Ustilago avenae*, the fungus that causes loose smut of oats. Although found everywhere oats are grown, the disease is rarely a serious problem because it is spread on infected oat grains and these can be easily treated to kill the fungus before the crop is planted.

This disease is not confined to oats and also occurs on some closely related grasses found in pastures. One such grass is *Arrhenatherum elatius* (tall oat grass), common in temperate regions of Europe and introduced to other places throughout the world. *Ustilago avenae* is conspicuously present on the ovaries and associated structures of the flower in *A. elatius*, where it takes the form of a small black, cigar-shaped mass.

When searching for smuts in the field, any plants in flower are examined for the presence of similar structures. Not all areas of the world have been intensively studied for smuts, so the number of described species will surely increase over time.

10

Interactions of Fungi and Animals

Leaf-cutter Ants 187
Fungus-cultivating Termites 188
Ambrosia Beetles 190
Wood Wasps 192
Yeast-feeding Bark Beetles 193
Fruit Flies, Soft Rots, and Cacti 195
Nematode-trapping Fungi 196
Free-living and Plant-Parasitic Forms 196
Trapping Devices 197
Adhesive Devices 197
Constricting and Nonconstricting Rings 198
Secretions 199
Fungivorous Flies 199
Mycophagous Mammals 200
Obligate Fungus Feeders 201
California Red-backed Vole 201
Australian Potoroos 202
Preferential Fungus Feeders 203
Northern Flying Squirrel 203
Red Squirrel 204
Spore Dispersal by Mammals 204

As described in Chapters 7 and 9, many different fungi form associations with plants or algae that are beneficial to both of the organisms involved. Mutualistic associations—two or more organisms living together such that both are more successful within the partnership than they would have been if they were living on their own—originated as a survival method at least 400 million years ago in the form of lichens. The two most important examples are mycorrhizal associations and the associations that exist between fungi and algae in the composite organisms known as lichens. Because both organisms benefit from the association, it represents an example of what is referred to as mutualism.

An obvious question to ask is whether or not fungi ever form mutualistic associations with animals, and the answer is a resounding yes. Two of the more unusual mutualistic associations that exist between animals and fungi involve the social insects. In both instances, the association is essential for the survival of the insects, for like animals, insects have never developed the enzyme systems needed to break down cellulose and lignin, the two most important components making up the tissues of vascular plants. Fungi, however, do possess the capability of breaking down cellulose and lignin. Leaf-cutter ants and mound-building termites represent the two best-known examples of fungus-animal mutualistic associations.

Leaf-cutter Ants

The leaf-cutter ants belong to a group known as attines that includes the genus *Atta*. These ants occur in the New World between latitudes 40° N and 44° S but are most abundant in the tropics of Central and South America. Attines are known for their extraordinary ability to defoliate vast tracts of forest in a matter of days. A mature colony can contain more than eight million individual ants in a nest half the size of a football field.

All the ants in the colony are specialists—the queens that are the breeders and also start the "fungus gardens" that will be described below, the ants that specialize in cutting the leaves from the trees, the ants that specialize in reducing the leaves to a mulch, and the ants that specialize in harvesting the fungi. There are also garbage worker ants that help manage the waste in garbage chambers and are kept isolated from the rest of the colony.

The attines are known to include 210 species within 12 genera that have formed mutualistic associations with fungi. At least seven genera of fungi are involved, all of which seem to be white-spored agarics from the family Lepiotaceae. The mutualistic association between leaf-cutter ant and fungus is so absolute that neither is found in nature without the other.

In brief, the ants that specialize in cutting leaves from a tree exit the nest and forage for suitable leaves. They use their mandibles to cut out a portion of a leaf, which they then transport back to the nest (PLATE 105). As a result of their foraging activities, the ants create very noticeable paths or tracts through the vegetation. As they travel away from and then return to the nest, all of the ants moving in the same direction form a distinct line.

The movements of the ants, especially those carrying a piece of a leaf or other plant part, are quite intriguing, as I can attest after observing leaf-cutter ants on many occasions in Costa Rica. At the nest, ants in the first group give their leaf pieces to a second group. The latter first cut the leaf pieces into smaller (usually 1/32 to 3/32 inch [1 to 2 mm]) portions and then use their mandibles to crush these smaller portions into a pulp. Saliva and fecal droplets are added to the pulp, and this mixture is deposited into the fungus gardens. The newly added material is "inoculated" with bits of mycelium obtained from other parts of the garden. The mycelium permeates the mass of organic matter present in the garden, forming what is essentially a "pure culture" of the fungus in question.

The culture is maintained and portions of the mycelium eventually harvested and used as a food source by the ants in the colony. The ants are able to keep their cultures free from other fungi except for species of *Escovopsis*, an ascomycete (a member of the order Hypocreales) that is a highly specialized fungus garden parasite. Bacteria that are carried on the exoskeleton of garden-tending ants produce fungicidal substances that suppress the growth of *Escovopsis* but do not affect the cultivated fungus in the garden itself. If the ants or the bacteria they carry are removed from the garden, the culture becomes badly contaminated almost immediately.

Interestingly, leaf-cutter ants are sensitive enough to react to the growth of the mycelium on different types of plant material, apparently responding to some type of biochemical signal from the fungus. If a particular type of leaf is toxic to the fungus, the colony will no longer collect it.

Fungus-cultivating Termites

Termites are widely known as destroyers of wood, but the taxa involved actually depend upon protozoans and bacteria in an enlarged portion of their digestive tract to break down the cellulose component of the wood they ingest. Not all termites have this same capability and must derive their nourishment from another source. Surprisingly, about 75 percent of all species of termites fall in the latter category.

What these termites do is very similar to what was described for leaf-cutter ants. Like the ants, fungus-cultivating termites occur in large colonies, and each colony consists primarily of different types of workers that specialize in performing certain tasks. In some species, the nest for the colony is a large moundlike structure that often exceeds a height of 6 feet (2 m) (PLATE 106). These mounds are true engineering marvels, and the termites use amazingly sophisticated methods to regulate their internal environment. For example, each mound has an elaborate ventilation system that enables the interior to have carbon dioxide and humidity levels that exceed those of the outside air.

Within the mound, the termites construct numerous structures called fungus combs. These consist of macerated plant material. The first step in constructing a comb involves specialized worker termites that forage outside the nest for wood and other types of plant debris, which they chew up and swallow. Upon returning to the nest, the workers deposit the macerated plant material as fecal pellets. Other worker termites who remain in the nest take these pellets and use them to construct the comb. The structure becomes inoculated with the fungus that is being maintained by the termites, and a mycelium develops and then spreads throughout the comb. As this happens, the plant material that makes up what is now a fungus comb is converted into a food source that can be utilized by the termites.

The total mass of fungus combs in an active termite colony is impressive. Typically it is eight times the mass of all the termites present in the colony, and a large colony is estimated to contain more than 90 pounds (40 kg) of fungus combs.

Fungus-cultivating termites are found in tropical Africa and Asia, and two of the more widely distributed genera are *Microtermes* and *Macrotermes*. Species of the latter construct the moundlike nests described above.

Species of *Termitomyces* (termite fungus) are the fungi that most commonly form an association with termites. This genus is a pink-spored member of the agaricales. The mycelium does not produce fruiting bodies while the nest is active, but once it has been abandoned, fruiting does occur. This takes place during the early part of the rainy season. Because the mycelium is well inside the mound, the fruiting bodies must break through the hard soil-like material of which the mound is constructed.

The fruiting bodies of *Termitomyces* are remarkable for their long stalks, each with a tapering rootlike extension called a pseudorhiza, and their sheer overall size (PLATE 107). *Termitomyces titanicus*, a species found in Zambia, can sometimes produce a fruiting body with a cap more than 2 feet (0.6 m) in diameter.

Species of *Termitomyces* are edible, and considerable numbers of fruiting bodies are collected each year. These can be purchased in local markets and roadside stands in the parts of Africa where the fungus occurs.

Ambrosia Beetles

The association between certain fungi and ambrosia beetles is not quite as sophisticated as the two examples described above. In fact, these beetles usually go unnoticed in the forests where they occur, but their activities are ecologically important to the forest as a whole.

The term "ambrosia beetle" does not denote a particular taxonomic group and actually encompasses at least 3000 species in two different subfamilies (Scolytinae and Playpodinae) of the family Curculionidae (weevils) that are obligately dependent upon fungi. These beetles all excavate tunnels in dead or dying trees, introduce their associated "ambrosia fungus" to the host tree, and then feed upon the mycelium of the fungus growing on the tunnel walls. The fungus is their sole source of food.

The first scientists to study these beetles in the nineteenth century had no idea as to what they were feeding upon, since insects were known to be unable to digest wood. When it was determined that the beetles actually fed upon the white, crystalline layer found on the walls of their tunnels, the substance making up the layer was named "ambrosia" (literally "food of the gods") because its origin was not yet known (thus seemingly representing a food provided by the gods) and also because it had a honeylike taste.

The large numbers of species involved throughout the world suggest that the strategy employed by ambrosia beetles is very successful, and this certainly seems to be the case. Other species of beetles ("bark beetles") are found in the narrow zone just beneath the bark of a tree, where they feed upon living tissue (especially phloem) but do not cultivate fungi. In the world as a whole, ambrosia beetles far outnumber bark beetles.

Of the two subfamilies of ambrosia beetles, members of the Scolytinae are comparatively rare in temperate forests but common in tropical forests, whereas just the reverse is true for members of the Playpodinae. Species of adult beetles in the two subfamilies differ greatly in their appearance, but their larvae and habits are very similar. The adult beetles are small, usually no more than 1/4 inch (0.5 cm) long. Upon locating a suitable host tree, the beetles bore directly into the wood and produce masses of very fine, dustlike particles of wood (borings) that are pushed out of the tree. The borings accumulate in bark crevices beneath the entry hole or fall to the ground below. The beetles usually select trees that are dead, dying, or weakened by disease, drought, or some other stress factor. Such trees are preferred because the moisture content of living, healthy trees is too high to allow the fungal associate to become easily established. Most species of ambrosia beetles occur on a wide range of hosts, but others display varying degrees of host specificity.

The tunnels excavated by the beetles can form a rather intricate network that penetrates some distance into the wood of the tree. Because the beetles themselves are small, the tunnels (usually called galleries) they excavate are usually no more than 1/8 inch (3 mm) in diameter, and the term "pinworm" has been used to refer to the beetles responsible for the tunnels when the latter are observed to be present in wood. Such tunnels are commonly encountered when splitting firewood.

Once the system of tunnels has been excavated, the host tree is "inoculated" with the fungus that will eventually infect the wood, forming "fungus gardens" for the beetles. These fungus gardens consist of thin crystalline-like mats of white, yellow, gray to dark brown hyphae that line the inner walls of the tunnels. Projecting into the tunnel from each hyphal mat are numerous tightly packed conidiophores bearing conidia. These are formed only after both the adult and larval beetles begin feeding upon the mats of hyphae (PLATE 108). The fungus (probably as conidia) arrived along with the beetles when the latter discovered the new host tree.

The fungus is carried in the digestive tract of the beetle or in special pocketlike cavities (called mycetangia) in the exoskeleton of the beetle. Considerable variation exists with respect to the structure and location of mycetangia (singular: mycetangium) in different taxa of beetles. However, they all have the same function—to transport the fungus. In the Scolytinae, mycetangia are usually found only in female beetles, but in the Playpodinae they are present in male beetles.

After the "fungus gardens" have been established in the host tree, the beetles excavate more tunnels, creating an extensive system of galleries. In some members of the Scolytinae, larger "brood chambers" are excavated. The female beetles lay their eggs in the galleries. When the eggs hatch, the larvae feed upon the fungus, which is essentially maintained as a pure culture in much the same manner already noted for leaf-cutter ants. Once again, it is not clear how this is done, but once the beetles leave the tunnels, they become contaminated with other kinds of fungi. Eventually, when adult beetles disperse to locate new host trees, they take the fungus with them, to renew the cycle of infection with a new generation of beetles.

The basis for the ambrosia beetle–fungus association is that the wood of a dead tree, although providing a favorable site for the beetles, is not available to them as a food source. The fungus, however, is able to convert the wood into a rich and readily available food source upon which the beetles can feed. Clearly, the beetles benefit from this association, since they could not survive on their own. The fungus is not as dependent upon the association but does gain considerable benefit from it. This would include being disseminated for some distance to new potential hosts by the beetles and then introduced directly onto substrates that are especially suitable for the growth and development of the fungus in question.

The taxonomy of the fungi actually involved in ambrosia beetle–fungus associations is not yet completely known. The majority appear to be anamorphs of various taxa in the ascomycetes. Many of these have been placed in the genus *Ambrosiella*, with others assigned to such genera as *Monacrosporium*, *Phialophoropsis*, and *Raffaelea*.

Wood Wasps

The insect-fungus association that exists for some types of wood wasps (especially species of *Sirex*) is quite different from the one described for ambrosia beetles, although the very same tree can be involved. Wood wasps generally attack conifers that are dead or dying. The female wood wasp is equipped with a long ovipositor that she uses to deposit her eggs in the wood of the tree, sometimes to a depth of an inch (2.5

cm) or more. The eggs hatch into larvae that tunnel into the wood and feed upon the mycelium of a wood-decaying fungus found on the tunnel walls. Eventually, the larvae develop into pupae and then adults, and the latter leave the tunnels to fly about for the rest of their lives.

The presence of a wood-decay fungus in the wood through which the larvae bore their tunnels is not something left to chance. When the female wasp deposits her eggs, she also inoculates the wood with the fungus. To do this, she carries short fragments of fungal hyphae (called oidia) in two specialized saclike structures in her abdomen. These structures (mycetangia), which literally serve as incubation chambers for the fungus, are connected to the anterior end of the ovipositor. During egg-laying, the mycetangia contract and some of the oidia (singular: oidium) are squeezed out and deposited into the wood along with the eggs.

The fungi involved are species of *Amylostereum* or *Stereum*, two basidiomycetes that cause a white rot of the wood. The fungal inoculum carried by the female wasp is acquired while she is still a larva and is retained as the pupa develops into the adult insect. Although this association is not as strictly defined as is the case for ambrosia beetles, the ultimate benefits are much the same for the fungus and the wasp. The fungus benefits from being dispersed to new suitable hosts, and the wasp benefits from having a more readily available food source than would otherwise be the case (Gilbertson 1984).

Yeast-feeding Bark Beetles

Although most bark beetles do not cultivate fungi in the manner described for ambrosia beetles, this does not mean that they are not associated with fungi. Many species of beetles appear to have a feeding strategy that involves yeasts or other fungi related to those associated with ambrosia beetles. When the bodies of field-collected bark beetles are examined, yeasts often can be found on the external surface of the exoskeleton or in the digestive tract, but they more commonly occur in specialized mycetangia. The latter can range from simple pits to more elaborate structures, and their presence is a clear indication that the beetles can serve as vectors for yeasts.

Most species of bark beetles attack trees that are weakened or dying because of disease, drought, smog, or physical damage. When they reach a new host tree, the beetles introduce the yeasts into the living tissue (phloem) just beneath the bark. The yeasts quickly spread throughout the phloem and ferment the sugars present. The bark beetles then feed upon the infected phloem. Some of the compounds produced by the fermenting yeasts appear to be involved in the chemical communication displayed by bark beetles.

Many species of bark beetles are colonial, and the effect of these compounds is to attract numerous other beetles of the same species to the host tree. The arrival and subsequent activities of these beetles are detrimental to the tree, often killing it. In some instances, a serious outbreak of the beetles in a particular forest can be ecologically and economically significant. When adult beetles disperse to new host trees, they carry yeasts with them. The yeasts involved in these associations include a number of different taxa, and the relative importance of the association to bark beetles appears to vary from one species of beetle to another.

Bark beetles are not the only insects known to have associations with yeasts. It has become increasingly apparent that yeasts are commonly present in the digestive tracts of numerous insects associated with plants and fungi. These insects include many different kinds of plant and fungus beetles along with bees and wasps, flies, earwigs, lacewings, crickets, and roaches. Although the nature of the insect-yeast relationship is not completely understood, it seems likely that the yeasts play an important role in detoxifying certain types of organic material consumed by their insect hosts while also supplying the latter with essential nutrients (Suh et al. 2008).

Meredith Blackwell, a mycologist at Louisiana State University, and several of her colleagues have discovered that the assemblage of yeasts found in the digestive tracts of insects that consume the fruiting bodies of fungi contains an astonishing diversity of undescribed species. In one study, carried out over three years, at least 200 apparently undescribed species of yeasts were isolated from a variety of different beetles collected in several localities in the southeastern United States and on Barro Colorado Island in Panama (Suh et al. 2005). These results suggest that significant numbers of yeasts are still unknown to science. Moreover, they provide abundant evidence that insect-yeast associations are both exceedingly common and widespread in nature.

Fruit Flies, Soft Rots, and Cacti

Yeasts have long been known to represent an important food item for a variety of flies. Among these are members of the genus *Drosophila* (fruit flies), which are characteristically found around overripe or rotting fruit and represent the most common small flies likely to be noticed in an ordinary kitchen. One species (*D. melanogaster*) has been used extensively for research in genetics and is a common model organism in developmental biology. Indeed, the terms "fruit fly" and "*Drosophila*" are often used synonymously with *D. melanogaster* in modern biological literature, although that species is just one of at least 1500 in the genus.

Drosophila larvae feed almost exclusively on yeasts, and for some species of both flies and yeasts, there is more involved than just a chance encounter in nature. In desert areas of the southwestern United States where cacti are found, a group of *Drosophila* inhabits cacti. These cactophilic ("cactus loving") flies lay their eggs and feed in soft rots that develop on portions of a cactus. The species of cacti involved include the giant columnar saguaro cactus (*Carnegia gigantea*) that occurs in portions of the Sonoran Desert, including southern Arizona. When a soft rot develops on a cactus, which can happen in response to physical damage or disease, the tissues involved are soon visited by cactophilic *Drosophila* that bring yeasts with them. Once established in the soft rot, the yeasts provide a source of food for the flies.

The yeasts associated with soft rots are mostly specialists and the assemblage of species present is distinct from those associated with other parts (for example, flowers or fruits) of the same cactus plant. *Drosophila* larvae are capable of distinguishing among the kinds of yeasts present in their immediate environment and spend more time feeding upon the preferred species. Moreover, adult flies have been shown to display feeding preferences that are different from those of larvae of the same species.

The association between yeasts and cactophilic flies is not as absolute as the associations that exist for the fungi maintained by ants, termites, and beetles. The yeasts are perfectly capable of surviving on their own, and more than a single species of yeast can serve as a food source for the flies. Moreover, the soft rot microhabitat where yeasts and flies coexist can be viewed as little more than a temporary "window of opportunity" for the flies involved and not something that requires the highly organized behavior demonstrated by fungus-cultivating ants and termites or even the somewhat less sophisticated behavior involved in the interactions of ambrosia beetles and their fungal associates.

Nematode-trapping Fungi

In each of the fungus-animal associations discussed thus far, both of the organisms involved benefit. Such is not the case for the association between nematode-trapping fungi and the animals they trap. The fungus certainly benefits, but the animal definitely does not.

The nematode-trapping (or nematophagous) fungi constitute a diverse group both morphologically and taxonomically. Most are anamorphs of taxa assigned to the ascomycetes, but a few examples belong to the zygomycetes or the basidiomycetes. All of the fungi involved tend to share the common feature of being saprotrophs typically associated with substrates low in available nitrogen.

The body of a nematode, or any other animal, contains abundant nitrogen, and the nematode-trapping fungi have simply evolved various ways of obtaining this nitrogen (albeit at the expense of the nematode). The situation is not unlike the one that exists for such insectivorous plants as the Venus flytrap (*Dionaea muscipula*) or the sundew (*Drosera*), which occur in habitats where the soils are deficient in nitrogen. These plants trap small insects to supplement the nitrogen available to them from the soil.

The fact that nematodes represent the "target" organisms for certain fungi is not surprising. Nematodes are among the most numerous animals on the earth. More than 80,000 species have been described, but the actual number of species is certainly much higher.

Free-living and Plant-Parasitic Forms

Free-living nematodes are common in most terrestrial habitats, and they also occur in both fresh water and salt water. These nematodes are particularly abundant in soil (especially soil rich in organic matter), dung, and decaying plant material, where many species feed upon bacteria or (less commonly) fungal hyphae. In addition to the free-living forms, many nematodes are parasites of animals or plants. The plant-parasitic nematodes are of considerable economic and ecological importance, and nematodes occur as internal parasites of most animals (including man).

It has been said that if the world (apart from nematodes) were to disappear, the ghostly forms of everything (land surfaces, bodies of water, plants, and animals) that had been present would be recognizable from the populations of nematodes. In short, a plentiful supply of nematodes is available in any habitats in which fungi are likely to occur, so the problem for the fungus is simply one of finding a way to take advantage of this situation.

Trapping Devices

Nematode-trapping fungi use a number of different devices to capture their prey. Six fundamentally different types of devices are generally recognized, and these can be broken down into adhesive and non-adhesive groups. Fungi in the first category produce a type of sticky secretion (essentially a "super glue" for nematodes) to which the body of a nematode readily and strongly adheres. Those in the second category make use of some non-adhesive device, a number of which are quite sophisticated.

Adhesive Devices

The simplest trapping device consists of adhesive side branches or specialized adhesive cells (called knobs) that arise from otherwise non-adhesive hyphae that make up the bulk of the mycelium of the fungus. These branches and cells occur at intervals along a given hyphae, so that the body of a nematode has a good chance of becoming attached to the mycelium at a number of places. If this is not the case, there is an excellent chance that the nematode will escape, since the response of a nematode upon making contact with one of the adhesive devices is to thrash violently about.

In other species of nematode-trapping fungi, the specialized adhesive knobs break off and are carried away by the nematode. Although it would appear that the nematode has managed to escape, this is hardly the case. The adhesive knob remains firmly attached to the nematode and soon gives rise to a hypha that penetrates the body wall, beginning a process that ends with the body of the nematode being riddled with hyphae. Since the nematode can travel some distance from the time it acquires the adhesive knob to when it finally succumbs to the infection by the fungus, the adhesive knob also provides a means of dispersal for the latter.

The most common type of trapping device consists of a network of interconnected hyphae and is known as an adhesive net. This may take the form of a ladderlike arrangement of simple hoops or a complex three-dimensional structure. The entire surface of the network is covered with the sticky secretion. Because adhesive nets are relatively extensive, they tend to be more effective in immobilizing a nematode. *Arthrobotrys oligospora*, considered the single most common species of nematode-trapping fungus, produces an adhesive net (PLATE 109).

Constricting and Nonconstricting Rings
The most sophisticated trapping devices are two different types of specialized three-celled rings that arise as lateral branches from vegetative hyphae. The hypha making up the lateral branch grows around in a perfect circle, finally fusing with itself just above the place where it originated. The end result is a three-celled, ring-shaped structure with a short stalk. The inside of the ring is about the same diameter as the body of a typical nematode. As a result, when a nematode happens to enter a ring, the latter fits snugly around the nematode.

There are two types of rings produced by various taxa of nematode-trapping fungi, and each works in a different manner. In non-constricting rings, the stalk is weaker than the ring itself, and the ring is easily detached. The nematode continues on its way, bearing a newly acquired collar around its body. When the fungi that produce this type of ring are common, it is not unusual for a particular nematode to have several collars present. What happens next is the same process of hyphal penetration and proliferation throughout the body of the nematode described above.

Constricting rings are similar in appearance to non-constricting rings, but the stalk of the ring is much stronger, and the ring is not easily detached. When a nematode enters a constricting ring and comes into contact with its inner surface, all three cells inflate rapidly, constricting the body of the nematode. A constricting ring can be considered as a "snare" and it almost always retains the nematode until it can be invaded by the hyphae of the ring-producing fungus.

Secretions

The dead wood of logs and stumps is low in available nitrogen, and two genera (*Hohenbuehelia* and *Pleurotus*) of wood-decaying basidiomycetes have evolved the capability to capture the nematodes that are common in woody substrates. In species of *Hohenbuehelia*, specialized hyphae of the fungus produce short, hourglass-shaped knobs that are enclosed by a droplet of a sticky secretion. Nematodes making contact with the knobs are held firmly in place until the hyphae of the fungus can invade their bodies.

In *Pleurotus*, specialized hyphae secrete droplets of a toxin-containing substance that paralyzes nematodes upon contact (Hibbett and Thorn 1994). The toxin works very quickly, so that affected nematodes have no opportunity to travel very far. Vegetative hyphae of *Pleurotus* grow to and then invade the bodies of the nematodes. That *Pleurotus* is a "predatory" fungus might seem somewhat surprising since it is commonly collected for human consumption. Presumably, this is one of the instances in which "you are what you eat" does not apply.

Fungivorous Flies

Anyone who has ever collected fleshy fungi in the field has probably noticed that some fruiting bodies are "wormy" when taken apart, a condition that appears to be especially common in boletes. Such fruiting bodies typically have a stalk and cap perforated with little tunnels inhabited by what might appear to be worms but are actually larval flies or (less often) beetles (Bruns 1984). These mycophagous insects have received much less study than their relative abundance in nature would seem to warrant.

One of the first things that becomes apparent in this association of insects and fungi is that the insects involved usually demonstrate a preference for fruiting bodies in a particular condition or stage of development. This is particularly evident in some of the fungivorous ("fungus feeding") flies. Some examples (known as primary fungivores) are attracted to young fruiting bodies, where the females deposit their eggs. The eggs quickly hatch and the larvae that emerge feed directly upon newly developed tissues. In other examples (known as secondary fungivores), the flies are attracted to and deposit their eggs in older, already decaying fruiting bodies.

Members of the families Anthomyiidae, Mycetophilidae, and Phoridae are the most frequently encountered primary fungivores and thus are responsible for the majority of "wormy" fruiting bodies. After a period of feeding, the fly larvae develop into pupae and then adults. The latter emerge and disperse, sometimes several months after the egg from which they developed was deposited. If the fruiting body of a bolete or other fleshy fungus is placed in some type of rearing chamber and the latter maintained for several weeks or months, adult flies can be collected as they emerge.

A simple rearing chamber can be constructed by placing a small amount of sawdust in the bottom of a glass jar and then covering the top of the jar with a small piece of a thin cloth or mosquito netting secured with a rubber band. I have had several students carry out undergraduate research projects that made use of such rearing chambers, and the results they obtained certainly suggested that fungivorous flies are amazingly common in nature.

Mycophagous Mammals

Many different mammals may consume fungi (Claridge and May 1994). The degree to which they depend upon fungi varies considerably. If fungi represent a significant part of their diet, the term "mycophagous" (meaning "fungus feeder") is appropriate to use. For some mammals, virtually the entire diet consists of fungi, and these would be considered as obligate fungus feeders. Certain other mammals typically feed upon fungi when these are available but switch to some other source of food when they are not. These would be considered as preferential fungus feeders. In other instances, a particular mammal will feed upon fungi if provided with an opportunity to do so, but fungi are not an essential part of its diet. Such mammals would be considered as opportunistic fungus feeders. Most mammals are not mycophagous at all but might on rare occasions ingest fungi inadvertently while consuming some other type of food. In this case, the mammal is no more than an accidental fungus feeder. Presumably, humans who go out collecting fungi for the table would be placed in the opportunistic category, while the vast majority of Americans would warrant no more than inclusion in the accidental category.

Obligate Fungus Feeders

Only a few mammals worldwide derive all, or nearly all, of their nutrition from fungi. Obligate mycophagy places some very severe constraints on the mammal involved. The most important of these is that the mammal must live in habitats where the fruiting bodies of fungi are available throughout the entire year or, if this is not the case, where there is some other alternative source of food that the mammal can utilize.

The fruiting bodies of most macrofungi (for example, most agarics and boletes) are relatively ephemeral or, if they do persist for a reasonable period of time, as is the case for most of the perennial polypores, they are tough or woody. In most regions of the world, it is possible to go out in nature during certain times of the year or after a period of very dry weather and be hard pressed to find any fruiting bodies that might look as if they would be palatable for a small mammal.

In one group of fungi the fruiting bodies remain in good condition for a reasonable period of time and are produced in situations where environmental conditions such as the amount of rainfall are much less of a factor. In this group of fungi the fruiting bodies are found underground. These hypogeous fungi (truffles and similar forms) are not uncommon in some forest ecosystems. Since the fungi involved form ectomycorrhizal associations with certain taxa of forest trees, only those forests in which these trees are present would represent a suitable habitat for a mammal that was an obligate mycophagist. Additionally, because truffles have to be located and then extracted from the ground before they can be consumed, such a mammal must be able to perform such tasks. Obligate mycophagists invariably have a keen sense of smell, which they use to find a truffle, and most have well-developed "digging" claws on their forefeet that allow the mammal to more efficiently extract it from the ground (Maser et al. 2008).

California Red-backed Vole

The California red-backed vole (*Myodes californicus*) is probably the best example of an obligate mycophagist in North America. This small mammal, which has a total body weight of no more than about an ounce (30 grams), depends almost entirely on truffles as a source of food. The voles spend most of their lives in or beneath the forest floor litter layer in old-growth forests of the Pacific Northwest. These forests remain relatively moist throughout much of the year and have an abundance of coarse woody debris, including well-decayed logs and stumps, on the forest floor. Such conditions are exceedingly favorable for both voles and truffles.

Analyses of the stomach contents of voles collected in the Coast Range of Oregon indicated that truffles made up more than 85 percent of the food items present, with little or no seasonal fluctuation (Maser et al. 2008). Lichens represented the only other food item of note and apparently served as an alternative source of food in late winter when the fruiting bodies of truffles would have been least abundant. Comparable data obtained in other parts of the vole's range indicated that an increased reliance upon lichens was apparent in places where winters were more severe and the fruiting season for truffles correspondingly shorter. Voles feed upon a wide variety of different fungi, and in one study the spores of nineteen different genera were identified from an examination of vole fecal pellets.

Australian Potoroos

There appear to be relatively more obligate mycophagists in the forests of Australia. One example is the long-footed potoroo (*Potorous longipes*), a housecat-sized marsupial found in the warm-temperate rain forests of eastern Victoria and New South Wales (PLATE 110). Surprisingly, the long-footed potoroo was unknown to science until 1980, when the first specimens (including one as a road-kill) became available. Studies of the animal carried out over the next two decades quickly confirmed its obligatory dependence upon truffles. One such study, carried out over thirteen months, revealed that the proportion of fungi in the diet never dropped below 80 percent.

Because of their dependence upon truffles, long-footed potoroos are limited to habitats where these fungi are predictably available throughout the year. Few areas in southeastern Australia are moist enough to support consistently high levels of truffle production, so relatively little suitable habitat potentially exists.

Even in those habitats where the animal is known to occur, its distribution changes throughout the year. During the warm, dry summer months, long-footed potoroos are most likely to be found in riparian areas along streams, while during the cool, moist winter months, they move out into other areas of forests. It seems very likely that these changes in their distribution reflect changes in the abundance of truffles. Truffles appear to be especially diverse in the forests of Australia, and studies of fecal pellets from long-footed potoroos indicate that it feeds upon many different species, apparently including some that are still unknown to science.

Another example of an obligate mycophagist from Australia is the critically endangered Gilbert's potoroo (*Potorous gilbertii*) that is restricted to a very limited area in the southwestern corner of Western Australia. This animal had been presumed extinct for more than 120 years until a specimen was accidently live-trapped in 1994. The limited data available from the few specimens that have been studied indicate that they feed almost exclusively upon truffles.

Preferential Fungus Feeders

Several other species of Australian mammals are preferential mycophagists. For some of these, fungi may make up 80 to 90 percent of the total diet at certain times of the year. Prominent examples are the long-nosed potoroo (*Potorous tridactylus*), the brush-tailed bettong (*Bettongia penicillata*), Tasmanian bettong (*Bettongia gaimardi*), and the northern bettong (*Bettongia tropica*). These four species occur over a wide variety of different habitats that range from open, eucalypt woodlands to tropical rain forests. The one feature that all of these habitats share in common is the presence of trees that form mycorrhizae with the truffles upon which the animals feed.

Northern Flying Squirrel

The most widely distributed preferential mycophagist in North America is the northern flying squirrel (*Glaucomys sabrinus*), which occurs from the treeline in Alaska and Canada southward in the west to northern California and Colorado, in the middle of the continent to central Michigan and Wisconsin, and in the east to northern North Carolina and Tennessee. Throughout its entire range, this animal has been repeatedly demonstrated to feed upon truffles during portions of the year when these fungi are available. When truffles are not available, such as during the winter months, the northern flying squirrel makes the switch to other sources of food, which often include various lichens.

This squirrel is also known to cache food items to be consumed at some later time. These caches can be in cavities in trees, as well as in the squirrel's nest. Lichens and seeds are items most commonly cached.

Red Squirrel

Caching of fungi by red squirrels (*Tamiasciurus hudsonicus*) is well known, and similar behavior has been documented for a number of other small mammals, including Abert's squirrel and the European red squirrel. Red squirrels actually dry the fungi to be cached by hanging the fruiting bodies among the branches of trees. Numerous fruiting bodies are sometimes present in a single tree, as I have observed on several occasions during the course of carrying out field work in Alaska. My reaction upon seeing a "cache tree" for the first time was one of surprise. I already knew that squirrels exhibited this type of behavior, but actually observing it first hand was something else.

After the fruiting bodies have been dried, which usually takes no more than a couple of days, the squirrel stores them in knot holes, hollow branches, and nests of twigs high in the canopy, as well as in holes beneath logs and stumps at ground level. In Alaska, mycologist Gary Laursen and two of his colleagues (Laursen et al. 2003) reported that red squirrels sometimes cache fungi in the "witches brooms" caused by mistletoe infection of spruce trees. On rare occasions, fungi have been placed in old birds nests. In one truly extraordinary observation, Dice (1921) noted that a red squirrel had packed dried mushrooms into empty cans in an abandoned cabin in Alaska after first drying them on shelves and other surfaces. Most of the fruiting bodies cached by red squirrels are those of epigeous (above ground) species; relatively fewer records of caches of hypogeous ("trufflelike") fungi have been reported.

Spore Dispersal by Mammals

Mycophagous mammals obviously benefit from using the fruiting bodies of truffles as a food source. However, it would be a mistake not to consider some of the other aspects involved in the relationship that exists between the two organisms. The most important of these is the role that mammals play in the dispersal of the fungus.

When a mammal excavates and consumes a truffle, it ingests fungal spores. These spores are then deposited via fecal pellets in a new location as a result of the movements of the mammal. This has the effect of maintaining truffle diversity in a particular area of forest. In turn, truffle diversity is an important factor in maintaining the diversity of the host trees with which the truffles form mycorrhizae. When some type of disturbance (for example, fire) removes or dramatically changes the composition of the vegetation in a given area, the assemblage of truffles in that area may be depleted or even lost. The reintroduction of truffles often takes place when spores are deposited in the area through the activities of mycophagous mammals.

These same mammals are subject to predation by owls and other raptors. When captured and killed, the body of the mammal, along with the fungal spores in its digestive tract, can be transported to distances well beyond what the mammal would have traveled as a result of its own activities. The consequences of this type of spore dispersal are difficult to determine, but they should never be discounted.

By disrupting the soil when they excavate truffles, mycophagous mammals create small pits that allow water to percolate into the ground. The pits might not seem important, but they occur in places where the mycelium of the truffle-producing fungus is present and would theoretically benefit from an increased availability of moisture. Making water more available is probably very important in those areas of the world where the nature of the soil surface limits the infiltration of water as is the case in Australia, where eucalypt forests tend to have soils that are relatively impermeable to water.

11
Fungi and Humans

Bioluminescent Fungi 207

Fairy Ring Fungi 208

In Folklore and Shakespeare 209

In Nature and Landscapes 209

Fungi in Fantasy Literature 210

A Journey to the Center of the Earth 210

Alice in Wonderland 211

The First Men in the Moon 212

The Chestnut Blight Fungus in North American Forests 213

Fungi as Dyes 214

Litmus Paper 215

Harris Tweeds 215

Skin Stain 215

Wool Clothing 216

More About Dyes 216

Edible Fungi 216

Wild Fungi in Ancient Cultures 217

Commercial Cultivation of Shiitake 217

Fungus-fearing and Fungus-loving Societies 218

Medicinal Fungi 219

Treating Wounds 220

Combating Infection 220

Panaceas 221

Hallucinogenic Fungi 221

Little Brown Mushrooms 222

Fly Agaric 223

Fungi in Fermentation 224

Bread 224

Alcoholic Beverages 225

Mushrooms and other larger and more conspicuous fungi occur throughout the world in every type of ecological community from an ordinary urban lawn to a lush tropical rainforest. Presumably, humans have always been aware of these fungi, since they tend to be such a conspicuous part of nature. In some societies, there was probably little more just than a general awareness, while in other societies, mushrooms were consumed as food or used in some other fashion.

When the well-preserved natural mummy of a man from about 3300 BC was discovered in a glacier in the Italian Alps in 1991, he possessed portions of the fruiting bodies of two different species of bracket fungi. One of these was a type of polypore (*Piptoporus betulinus*) known to have antibacterial properties and most likely used for medical purposes. The other was a type of tinder fungus (*Fomes fomentarius*) and was part of what appeared to be a complex fire starting kit.

Bioluminescent Fungi

Throughout human history, night has been a time of resting and sleep, since little work or labor can be carried out in the dark. This was certainly the case before the invention of electric lights that we take for granted today. Night has also often been associated with danger, evil, and mythical creatures. Almost all cultures have stories and legends warning of the dangers of the night. It is therefore not too surprising that the mysterious lights sometimes observed at night in forests would evoke fear and wonder.

Such lights are actually the result of a natural phenomenon that involves certain bioluminescent fungi. The light these fungi emit is often referred to as foxfire in the popular literature.

Bioluminescence (biologically produced light) is known for a diverse assemblage of organisms that includes bacteria, certain algae, and insects (the common firefly is the most familiar example) in addition to fungi. The emission of visible light by a living organism is the result of an enzyme-catalyzed (luciferase) reaction of molecular oxygen with a substrate (luciferin). The organic molecules involved in the different organisms capable of bioluminescence are not biochemically identical, which suggests that the bioluminescent systems they possess evolved independently.

Bioluminescent fungi are not especially common, but three examples are common in North America: *Armillaria mellea* (honey mushroom), *Omphalotus illudens* (jack-o-lantern mushroom), and *Panellus stipticus* (luminescent panellus). *Armillaria mellea* appears to be the most common source of foxfire. This particular species emits a bluish-green glow. The bioluminescence is confined to the mycelium, which occurs in decaying wood.

In *Omphalotus illudens*, the gills are the source of the bioluminescence. The fruiting bodies of *O. illudens* are relatively large (with caps up to 6 inches [15 cm] wide) and usually occur in clusters on decaying logs or at the base of stumps or standing dead trees. The bioluminescence they produce can be observed only in fresh specimens within which basidiospores are still forming.

Panellus stipticus occurs on logs, stumps, and fallen branches in broadleaf forests. The fruiting bodies are relatively small (usually less than an inch [2.5 cm] wide), dingy white to pale brown, and kidney- or shell-shaped. They glow in the dark when fresh, and the light they emit is often brighter than that of *Armillaria mellea* or *Omphalotus illudens*. In general, foxfire is a very low intensity light and is not apparent if any other source of light is nearby. It is best observed on overcast or moonless nights and after the eyes have become fully dark-adjusted for at least 10 to 20 minutes.

Throughout the world, more than 40 species of bioluminescent fungi have been described, and the phenomenon appears to be most common in several genera of white-spored agarics. Perhaps the most prominent example is the genus *Mycena* (Desjardin et al. 2007). More species are known from the tropics than temperate regions of the world (see Plate 111 for an example from Brazil).

Historical evidence of foxfire extends back well over two millennia. The Greek philosopher Aristotle noted the occurrence of a "cold fire" light that is almost surely a reference to a bioluminescent fungus, and the Roman naturalist Pliny the Elder mentioned luminous wood in olive groves. The Swedish historian Olaus Magnus wrote in 1652 that people in the far north of Scandinavia would place pieces of rotten oak bark at intervals when venturing into the forest and then find their way back by following the light. In a similar fashion, during World War I soldiers in the trenches sometimes attached fragments of wood with a bioluminescent fungus present to their helmets to keep from bumping into others in the dark.

Fairy Ring Fungi

Because mushrooms and other fungi are so common and widespread, it is not surprising that they sometimes appear, usually in the background, of paintings that depict some aspect of nature. The Baroque style of art that appeared in Italy in the late sixteenth century and especially the Dutch still life painters of the seventeenth century sometimes incorporated mushrooms in their scenes of nature.

Mushrooms appear much more prominently in British paintings of the Victorian era that represent a genre known as Victorian Fairy Paintings. These paintings, which depict scenes of the hidden-from-view world of fairies, elves, and other mythical creatures, were very popular in the second half of the nineteenth century and the very early part of the twentieth century. They still have a certain appeal even today. The association of fairies with fairy rings is a common theme, and the painting *A Fairy Ring* by Walter Jenks Morgan (1847–1924) represents an especially good example (PLATE 112).

In Folklore and Shakespeare

A considerable body of folklore is associated with fairy rings. In northern Europe, the widely held belief that mushrooms growing in a circle followed the path made by elves or fairies dancing in a ring dates back to at least the Middle Ages. Entering a fairy ring was supposed to be dangerous or at least something that would bring on bad luck. Fairy rings have long been featured in the works of European authors and playwrights. William Shakespeare alludes to them in both *A Midsummer Night's Dream* and *The Tempest*. For example, the latter (Act V, scene 1, lines 41–45) contains the passage:

> you demi-puppets that
> By moonshine do the green sour ringlets make,
> Whereof the ewe not bites; and you whose pastime
> Is to make midnight mushrooms, that rejoice
> To hear the solemn curfew;

In Nature and Landscapes

As we now know, the tendency for fruiting bodies (mushrooms) produced by the mycelium of a particular fungus to occur in what is more or less a circular pattern (or ring) is a naturally occurring phenomenon and has absolutely nothing to do with fairies. These so-called fairy rings are found in both open grassy places and in forests, but they are most obvious when they occur in grassy fields, lawns, and other similar settings (PLATE 113).

Marasmius oreades (fairy ring mushroom) is the best known of the fungi that frequently form fairy rings, but a number of other species can do the same thing. The slightly upturned caps of some of the mushrooms depicted in Morgan's painting (PLATE 112) appear to show gills that are faintly pink. This is the color of the gills in fruiting bodies of *Agaricus campestris* (meadow mushroom) that are not yet fully mature. In fact, another common name (pink bottom) refers to this feature. *Agaricus campestris* commonly occurs in grassy areas, where it often forms fairy rings. While I was an undergraduate student at Lynchburg College in Virginia, this fungus formed a fairy ring in the lawn just outside the biology building, thus attracting the attention of the biology majors (including me) who were taking a class in botany at the time.

Both *Marasmius oreades* and *Agaricus campestris* are saprotrophs, and their mycelia occur in the soil. When a new mycelium becomes established, it has a tendency to grow outwards in all directions. As it grows, the mycelium uses up all of the nutrients in the soil, so that the older central portion dies. The still-expanding outer portion, which forms what is more or less a ring, is still actively growing, and when fruiting occurs, the fruiting bodies are produced just behind this outermost growing edge of the mycelium, thus forming the fairy ring.

Measurements of the distance that a fairy ring expands outward indicate a growth rate of no more than an inch or two (2.5 to 5 cm) per year. As such, a large fairy ring is likely to be very old. Some examples are estimated to have been in place for more than a century.

Fungi in Fantasy Literature

Mushrooms have been mentioned in a number of notable works of fantasy. Two prominent examples are Lewis Carroll's *Alice's Adventures in Wonderland* (more commonly known as *Alice in Wonderland*) and Jules Verne's *A Journey to the Center of the Earth*.

A Journey to the Center of the Earth

In *A Journey to the Center of the Earth*, mushrooms are described as reaching gigantic proportions in the newly discovered region at the earth's center.

But at that moment my attention was drawn to an unexpected sight. At a distance of five hundred paces, at the turn of a high promontory, appeared a high, tufted, dense forest. It was composed of trees of moderate height, formed like umbrellas, with exact geometrical outlines. . . .

I hastened forward. I could not give any name to these singular creations. Were they some of the two hundred thousand species of vegetables known hitherto, and did they claim a place of their own in the lacustrine flora? No; when we arrived under their shade my surprise turned into admiration. There stood before me productions of earth, but of gigantic stature, which my uncle immediately named.

"It is only a forest of mushrooms," said he.

And he was right. Imagine the large development attained by these plants, which prefer a warm, moist climate. I knew that the *Lycoperdon giganteum* [now known as *Calvatia gigantea*] attains, according to Bulliard, a circumference of eight or nine feet [2.4 or 2.7 m]; but here were pale mushrooms, thirty to forty feet [9.1 to 12.2 m] high, and crowned with a cap of equal diameter.

Although Verne certainly would not have been aware of the fact, there was one ancient fungus that did reach an extraordinary size, albeit it would not have looked anything like the mushrooms described in his novel. The fossil evidence of this ancient fungus will be described in Chapter 12.

Alice in Wonderland

In *Alice in Wonderland*, the English author Charles Dodgson (who was far better known by his pseudonym Lewis Carroll) describes what happens when his main character encounters a mushroom with truly extraordinary properties.

Alice looked all round her at the flowers and the blades of grass, but she did not see anything that looked like the right thing to eat or drink under the circumstances. There was a large mushroom growing near her, about the same height as herself; and when she had looked under it, and on both sides of it, and behind it, it occurred to her that she might as well look and see what was on the top of it. She stretched herself up on tiptoe, and peeped over the edge of the mushroom, and her eyes immediately met those of a large caterpillar, that was sitting on the top with its arms folded, quietly smoking a long hookah, and taking not the smallest notice of her or of anything else.

Alice engages the caterpillar in conversation, and the latter informs her that one side of the mushroom will make her taller and the other side will make her shorter. She breaks off two pieces from the mushroom. One of these makes her shrink smaller than ever, while the other causes her neck to grow high into the trees. With some effort, Alice eventually brings herself back to her normal height. There is little question that what is being described could be attributed to the effects of a hallucinogenic mushroom, and these were certainly known in nineteenth-century England at the time (1865) this story was published.

The First Men in the Moon

In his science fiction novel *The First Men in the Moon*, published in 1901, British author H. G. Wells described fungi occurring on the moon. The novel tells the story of a journey to the moon undertaken by the two main characters. Once arriving on the surface, they promptly get lost. Growing hungry, the two earthmen sample a lunar fungus.

> The stuff was not unlike a terrestrial mushroom, only it was much laxer in texture, and, as one swallowed it, it warmed the throat. At first we experienced a mere mechanical satisfaction in eating; then our blood began to run warmer, and we tingled at the lips and fingers, and then new and slightly irrelevant ideas came bubbling up in our minds.
> "It's good," said I. "Infernally good! What a home for our surplus population! Our poor surplus population," and I broke off another large portion. It filled me with a curiously benevolent satisfaction that there was such good food in the moon. . . .
> I think I forgot the Selenites, the mooncalves, the lid, and the noises completely so soon as I had eaten that fungus.

Soon after ingesting the fungus, a hazy euphoric state overtakes the two, and they wander drunkenly, speaking gibberish. Interestingly, their response is not unlike what some individuals have been known to experience after consuming hallucinogenic mushrooms such as *Amanita muscaria*. It seems rather apparent, based on the three examples considered herein, that literature is not the place to look for "ordinary fungi."

The Chestnut Blight Fungus in North American Forests

As described in Chapter 2, several fungi and funguslike organisms have had an undeniable impact upon human history by affecting agriculturally important plants. In a number of other instances, fungi have impacted whole ecosystems, but because the consequences to humans were not as direct, these other fungi have generally received far less attention in the public media.

Perhaps the best-known example is *Cryphonectria parasitica* (chestnut blight fungus). This ascomycete, formerly known as *Endothia parasitica*, was accidently introduced to North America from China or Japan about 1900, on either imported chestnut lumber or living trees brought in as nursery stock. Both Chinese chestnut (*Castanea mollissima*) and Japanese chestnut (*C. crenata*) are native to eastern Asia and are relatively resistant to the fungus. They may become infected, but they usually do not die as is the case for the American chestnut (*C. dentata*).

In 1904, when Herman W. Merkel, a forester at the New York Zoological Park in the Bronx, found some dying American chestnut trees, he could hardly have known that this event would cause what has been described as the worst ecological disaster in forest history. As was soon discovered, *Cryphonectria parasitica* is a highly virulent pathogen of American chestnut. Once established, the mycelium effectively girdles the chestnut stems, thus killing the tree. Within an amazingly short time, the chestnut blight (sometimes called chestnut bark disease) spread throughout the entire natural range of American chestnut in eastern North America. By the late 1940s the species had been essentially eliminated from forests over this entire region.

Chestnut wood is full of tannins and relatively resistant to decay, and some long-dead trees were still standing as "gray ghosts" in forests several decades after they had died. For my master's degree at Virginia Tech, I studied post-blight forests to determine just what trees had replaced chestnut. At the time I carried out my field work (mostly during the summer of 1969), there were still a reasonable number of gray ghosts in the mountain forests of southwestern Virginia.

Prior to the blight, American chestnut is estimated to have made up 40 percent of the canopy trees in the forests of eastern North America. In fact, it has been said that at one time a squirrel could jump from Maine to Georgia, chestnut tree to chestnut tree, without ever touching the ground.

Chestnut wood is straight-grained, strong, and easy to saw and split. When still readily available, this wood was used for a variety of purposes, including furniture, split-rail fences, shingles, and the construction of homes and barns. Tannins were extracted from chestnut bark for tanning leather, and more than half of the tannin used by the American leather industry at the beginning of the twentieth century was derived from chestnut. Interestingly, because it is so resistant to decay, a considerable amount of chestnut wood has been reclaimed from old barns and other buildings to be refashioned into furniture and other items.

In addition to its uses by humans, the fruit of the American chestnut was a valuable food source for wildlife, ranging from larger animals such as deer and black bears to birds (turkeys, grouse, and quail) and small mammals (raccoons, squirrels, and chipmunks). Clearly, the virtual elimination of chestnut by the blight had a profound effect in many different ways.

Despite the devastation the blight caused to a chestnut tree, the root collar and root system of the tree are relatively more resistant, so a large number of small trees still survive as root sprouts. These seldom grow to be very large before they are infected by the blight.

Surviving chestnut trees are being bred for resistance to the blight, notably by the American Chestnut Foundation, which has the objective of eventually reintroducing a blight-resistant American chestnut to its original range. One can only hope that this takes place.

Fungi as Dyes

The natural color of sheep's wool is rather drab by anyone's standards, and most of the earliest humans who would have worn wool clothing would have accepted this fact without much if any thought. It was simply the way things were. With the passage of time, natural organic dyes to change the color of wool became increasingly widespread. Many of these dyes were derived from various plants and animals, but some were derived from fungi and (especially) lichens. The most common colors obtained from fungi and lichens are brown, yellow, and gray, but other colors such as red, orange, dark blue, purple, and black can be obtained from some examples (PLATE 114).

Litmus Paper

Two of the dyes (one purple and the other red) commonly used in Medieval Europe were obtained from the crushed and boiled thalli of the lichen *Roccella tinctoria*. Some lichen-derived dyes change color in the presence of acids or bases, as is the case for the dye present on the litmus paper so commonly used in high school chemistry laboratories. This dye, which is derived from species of *Roccella*, turns red in acid solutions and blue in alkaline solutions.

Harris Tweeds

When coal tar dyes were invented in the 1850s, the use of all-natural dyes started to decline. Only a few lichen-derived dyes are still used today. One of them, a brown dye obtained from *Parmelia omphalodes* and *P. saxatilis*, is still employed on hand-woven Harris tweeds from the Outer Hebrides of Scotland. The two lichen dyes (referred to as crottle) produce the distinctive deep red-brown to purple-brown and rusty orange colors associated with this type of cloth. In addition, one of the dyes (orcein) used in microbiology laboratories for staining the chromosomes of a cell is derived from a lichen (*Roccella tinctoria*).

Skin Stain

Numerous species of fungi have been tested for possible use as sources for dyes. Many produce disappointing and often unpredictable results, since the color of the fruiting body can be lost or greatly modified in the dyeing process. *Echinodontium tinctorium* (Indian paint fungus) is an exception. This fungus yields a rich orange-red dye that was used by Native Americans in the Pacific Northwest to stain their skin. The somewhat hoof-shaped fruiting bodies of this polypore are large at 4 to

10 inches (10 to 25 cm) across and occur on living conifers (Miller and Miller 2006).

Wool Clothing

Fruiting bodies of *Pisolithus tinctorius* (dye-maker's false puffball) can be used as a source of a rich golden-brown to black dye (PLATE 115). Although the club-shaped fruiting bodies, which are 2 to 8 inches (5 to 20 cm) tall and have an upper portion that contains numerous small egglike structures, resemble those of puffballs, *P. tinctorius* is actually one of the earthballs. This fungus essentially occurs worldwide, often in disturbed sites with poor soils. Other fungi that can be used as sources of dyes are described in the book *Macrofungi Associated with Oaks of Eastern North America* (Binion et al. 2008).

More About Dyes

Although the examples mentioned here indicate that natural dyes derived from lichens still have at least some economic value, the almost universal use of synthetic dyes has caused fungus-derived dyes to become relatively inconsequential. In some respects this trend is unfortunate, since there is little question that natural dyes (including but not restricted to lichens and fungi) are inherently safer to use and more environmentally sustainable than synthetic dyes.

All of the information that anyone would need to produce and then use dyes derived from lichens and fungi is available in a number of books (for example, Rice and Beebee 1980) and is often the subject of workshops held in conjunction with larger gatherings (for example, the national foray held each year) of amateur mycologists. Moreover, projects that display the results obtained from mushroom-derived dyes are not infrequently encountered in high school science fairs. Having served as a judge for such fairs on numerous occasions, I recall several examples of this type of project, all of which were exceptionally well done.

Edible Fungi

Because such fungi as rusts and smuts are virulent pathogens of some of the plants that have long been important as agricultural crops, human-fungal interactions have not always been positive. However, the use of certain species of fungi either directly as a source of food or indirectly in the preparation of certain types of food or beverages is something that has gone on for a very long time.

Indeed, in *Gorillas in the Mist* (1983), Diane Fossey noted that mountain gorillas (*Gorilla berengei*) in the wild sometimes consumed a certain type of bracket fungus. The fungus in question seemed to represent a special treat, based on the behavior of any gorilla that happened to find it. Lal Singh, director of the Himalayan Research Group in northwestern India, once related to me that wild rhesus monkeys (*Macaca mulatta*) collect and eat many of the morels that appear in early spring in the mountain valley where he grew up. There is no reason to believe that the very first humans would not have done much the same thing as these two fellow primates.

Wild Fungi in Ancient Cultures

Eating wild fungi is first reliably noted in Chinese writings from well over 2000 years ago (Aaronson 2000). The oldest indisputable reference to edible fungi outside of Asia appears in the writings of the Greek philosopher Theophrastus, who mentioned truffles (Dugan 2008). The Greeks and (even more so) the Romans collected and consumed a number of different fungi. Not all of the taxa involved can be determined with any degree of certainty, but some can be identified and certain others can be considered as strong possibilities (Houghton 1885). The first group would include edible species of *Amanita*, especially *A. caesarea* (appropriately known as Caesar's mushroom), *Boletus edulis* (king bolete), and certain puffballs, while members of the second category would encompass *Agaricus campestris* (meadow mushroom), *Coprinus comatus* (shaggy mane), *Lactarius deliciosus* (deceptive milky), and *Russula alutacea* (yellow-gilled russula). It is not surprising that all of these fungi are still consumed today.

Commercial Cultivation of Shiitake

Throughout the world, many different kinds of fungi are consumed, and some of these are grown commercially on a large scale. In addition to *Agaricus bisporus* (commercial mushroom), *Lentinula edodes* (shiitake), *Volvariella volvacea* (straw mushroom), *Flammulina velutipes* (enokitake), *Tremella fuciformis* (white jelly fungus), *Pholiota nameko* (namely), *Dictyophora indusiata* (bamboo mushroom), *Agaricus blazei* (portobello mushroom), various species of *Pleurotus* (oyster mushroom), and *Auricularia* (wood ear) are important cultivated fungi (Hall et al. 2003). For many of these, commercial production is centered largely in Asia, but shiitake is now grown in many other places. In fact, it now trails only *Agaricus bisporus* in terms of total world mushroom production.

The cultivation of shiitake originated in China about 1000 AD. The traditional method of growing this fungus, which occurs as a saprotroph on decaying wood, involves inoculating logs with the mycelium of the fungus. Oak logs are the most widely used, but other types of logs (for example, beech or poplar) will support good growth of shiitake. The logs are placed in a shady situation, and their moisture content closely monitored. Six to twelve months after inoculation, fruiting bodies of shiitake begin appearing along the sides of the log. After these have been collected, it is usually possible to induce another cycle of fruiting by soaking the logs. Fruiting cycles are sometimes repeated for several years, until the wood of the log is almost completely decomposed.

A more recent method involves growing shiitake on sterilized sawdust in a plastic bag. The advantage of this method is that the fungus can be grown indoors, where environmental conditions can be more easily controlled.

Fungus-fearing and Fungus-loving Societies

The large-scale cultivation of fungi is a relatively recent innovation, and for most of human history, the fungi consumed were collected in the wild. Societies differ markedly with respect to mycophagy (consumption of fungi). Some societies give little evidence that fungi ever represented a significant food item, while in other societies fungi appear to be (and probably always have been) rather important. Any visit to a local market in Southeast Asia during the appropriate season will invariably reveal an enormous variety of different fungi for sale. It is interesting to note that there have been a few occasions when a mycologist, upon visiting such a market, has "discovered" a fungus previously unknown to science.

Fungi are not as prominent in markets throughout much of Europe and are often conspicuously absent from markets in Australia and the United States. Wasson and Wasson (1957) discussed the idea that a given society could be classified as mycophobic ("fungus fearing") or mycophilic ("fungus loving") based on the presence or absence of a general prejudice against collecting and consuming fungi. Although such an approach is too simplistic to be universally accepted by mycologists, clearly the differences they denote can and do exist.

Medicinal Fungi

In some societies, certain fungi have been valued not only as food but also for their perceived medical properties. This was particularly the case in Asia, where fungi have been used in traditional Chinese medicine for thousands of years (Hobbs 1995). One of the best known of these medicinal fungi was (and still is) *Ganoderma lucidum* (known as *ling zhi* in China and *reishi* in Japan), which has been used to treat liver diseases such as hepatitis, high blood pressure, arthritis, insomnia, bronchitis, and asthma. In addition, *G. lucidum* has long been thought to have properties that might extend the life span while increasing vigor and vitality.

Some of the other commonly used fungi in traditional Chinese medicine have been mentioned in other contexts elsewhere in this book. These include the vegetable caterpillar fungus (*Cordyceps sinensis*), shiitake (*Lentinula edodes*), and turkey tail (*Trametes versicolor*). As noted in Chapter 4, Chinese distance runners attributed their improved performances to the use of the vegetable caterpillar (known in China as *dong chong zia cao*). This fungus has long been used as both a nutritional supplement and a general lung and kidney tonic.

In the past, shiitake was used for any bodily condition in which the immune system needed a boost, which included anything from the common cold to cancer. The turkey tail (known as *kawaratake* in Japan and *yun-zhi* in China) was considered a useful treatment for infections of the respiratory, urinary, and digestive tracts. Although regarding any fungus as a panacea for a wide variety of human aliments is contrary to what Western medicine now knows, it would be a mistake to dismiss the fungi involved in traditional Chinese medicine. Results from laboratory and human clinical studies carried out on some of these fungi have shown that they do contain a number of chemical compounds with antitumor and immunostimulating properties.

Treating Wounds

Medical uses of fungi in other parts of the world appear to have been much less extensive than was the case in Asia, but some of these uses probably date back many thousands of years. One of earliest examples is likely to have been the use of the fruiting bodies of several species of polypores for staunching and dressing wounds. For example, in *Fomes fomentarius* and *Piptoporus betulinus*, the interior of the fruiting body is rather fibrous and absorbent. This material was cut into pieces and then pounded to make it soft and pliable, at which point it could be applied to the wound. It is now known that many polypores have antibacterial properties, so this was an additional factor that entered into their use in healing wounds.

A number of puffballs, including species of *Calvatia*, were used to stanch the flow of blood from wounds. In some cases, the soft, interior portion (developing gleba) of an immature puffball was dried, ground into a fine powder, and then applied to a wound. In other cases, spores from a mature puffball were used in the same manner. Both are known to promote rapid clotting of the blood as well as for having some degree of antibacterial activity.

Combating Infection

The discovery by Alexander Fleming in 1928 that a substance (later named penicillin) produced by a species of *Penicillium* inhibited the growth of certain types of bacteria is considered one of the noteworthy events of the twentieth century. Unarguably, penicillin has been responsible for saving many human lives. As the examples noted above indicate, the use of the antibacterial properties of fungi actually predate Fleming's discovery. In fact, it is even possible that *Penicillium* (and thus penicillin) may have been involved in some instances. Beginning at least as long ago as in ancient Egypt, pieces of moldy bread (or other types of organic material colonized by microfungi) were sometimes applied to surface wounds as a way of combating infection. On the basis of what we now know, this actually makes sense.

Panaceas

Ganoderma lucidum, the best known and most widely used fungus in traditional Chinese medicine, had somewhat of a counterpart in Europe. Like *G. lucidum*, *Laricifomes officinalis* is a polypore, and it was reputed to be a panacea for a wide variety of human aliments. The species (known to the Greeks as *agarikon* and the Romans as *agaricum*) commonly occurs on larch, and references to its medical properties appear in the writing of the Greek physician Pedanius Dioscorides, who described the fungus as an effective treatment for aliments ranging from bruises to broken limbs as well as an antidote to poisons. He also described a number of ways in which it could be administered (for example, with or without wine or honey). Dioscorides was the author of *De Materia Medica*, a major work on herbal medicine that was still being used as a standard source of information on this subject more than 1500 years after it was written.

Hallucinogenic Fungi

No discussion of the fungi used for medical purposes would be complete without considering species with psychoactive compounds that, when consumed by humans (and animals), induce an altered state of consciousness during which there are false perceptions of sight, sound, taste, smell, or touch. Such fungi, usually referred to as hallucinogenic, have been known for at least 2000 years and probably a lot longer. In many instances, they have played a role in the religion or mythology of a particular society. For example, the discovery of mushroom sculptures among the ruins of ancient temples in Central and South America suggests that the native people used hallucinogenic fungi during religious ceremonies.

The Aztecs are known to have used the term "teonanacatl" (meaning "flesh of the gods") to describe hallucinogenic fungi which included *Psilocybe mexicana* (sacred mushroom) and several related species. Historians have suggested that Aztec spiritual leaders used these fungi to induce the altered state of consciousness required for them to communicate with their gods. The psychoactive compounds in species of *Psilocybe* are psilocybin and psilocin, and their fruiting bodies exhibit a characteristic blue-staining reaction when bruised upon being handled.

Little Brown Mushrooms

Psilocybe is a genus of perhaps 300 species, many of which occur as saprotrophs on the dung of large herbivores. One example is *P. cubensis* (magic mushroom), which is common on cow dung, sheep dung, or horse dung in warmer areas of the Northern Hemisphere. Although limited by climate in nature, this species can be cultivated indoors on suitable substrates, and it has been rated as a "moderately potent" hallucinogenic fungus (Stamets 1996). During the 1960s and 1970s, when experimentation with drugs became fashionable in the United States and elsewhere, the properties of *P. cubensis* were soon very well known. Indeed, a passing reference to a "little brown mushroom" likely involved this species. The fruiting bodies of *P. cubensis* are relatively larger than those of most other members of the genus and can reach a total height of 4 to 6 inches (10 to 15 cm).

Panaeolus foenisecii (lawn mower's mushroom) is another little brown mushroom in which psilocybin can be found (PLATE 116). Unlike *Psilocybe cubensis*, *Panaeolus foenisecii* is found throughout the Northern Hemisphere and in other regions of the world as well. As the common name indicates, this fungus typically occurs in lawns, where the relatively small fruiting bodies with their bell-shaped caps and long stalks tend to form small groups. Although it can be hallucinogenic, *P. foenisecii* is rarely collected because the fruiting bodies are so small. The mushroom may pose a danger to small children who are at the stage when "everything goes to the mouth" if they consume it when left unsupervised in a lawn where the fungus is present. Unfortunately, incidents involving *P. foenisecii* and small children have happened from time to time.

Household pets also have been known to consume fruiting bodies of *Panaeolus foenisecii*. I am aware of one such occasion when a dog belonging to a friend of mine, who just happened to be a mycologist, began displaying unusual behavior after having had an opportunity to roam around in the lawn. My friend was puzzled by the dog's reaction but allowed it to leave the house again. This time he followed and observed the dog going directly to a group of fruiting bodies of *P. foenise-*

cii and then starting to consume them. Apparently, the dog either liked the taste of the fungus or the sensation it produced.

Fly Agaric

As was mentioned in Chapter 6, *Amanita muscaria* (fly agaric) is one of the best known of all macrofungi, a fact that can be attributed almost entirely to its hallucinogenic properties (PLATE 117). The common name refers to the use of this fungus, if portions of the fruiting body are placed in a shallow plate or bowl containing milk, to attract flies. Upon consuming some of the milk, the flies are stunned by the psychoactive compounds derived from the fungal tissue and then drown in the milk. There is some question as to how well this actually works, but it is an interesting story.

The psychoactive compounds in *Amanita muscaria* are muscimol and ibotenic acid, and their concentrations vary considerably from fruiting body to fruiting body and from place to place. Some individuals, upon consuming the fungus, experience no real effects at all, while others become ill enough to require a trip to the hospital. Fatalities are extremely rare.

There is a very large body of information (and some controversy) relating to the use of *Amanita muscaria* by humans and the hallucinogenic efforts it produces. There seems little doubt that this fungus played a significant role in the rituals and religious ceremonies carried out in some ancient societies. Perhaps the most important work on this subject is the book *Soma: Divine Mushroom of Immortality* by Gordon Wasson (1968). Wasson advanced the idea that soma, a ritual drink of considerable importance among early Indo-Iranians and the later Vedic and Persian societies, was derived from *A. muscaria*.

John Allegro (1970), known for his scholarly work on the Dead Sea Scrolls, suggested that the use of this same fungus by the people of ancient Mesopotamia and adjacent regions of the Middle East was the underlying basis for the religious experiences recorded in writings from the period. There is also evidence that *Amanita muscaria* was used as a hallucinogen in ancient and medieval Europe. One story that may or may not have a basis in fact relates to the purported use of the fungus by the Viking warriors who then became known as berserkers. These were warriors who, upon going into battle, fought ferociously with seemingly no regard for pain or injury. Presumably, the effects produced by the psychoactive compounds present in *A. muscaria* could account for such behavior. Clearly, when these and other similar historical accounts are taken into consideration, there is little wonder that so much has been written about this one species.

Fungi in Fermentation

As far as fungi go, the fruiting bodies of *Amanita muscaria*, which can have a cap the size of a small dinner plate, are relatively large. Moreover, as described above, this species has been of considerable importance in human affairs. Nevertheless, there is another species among the smallest of all fungi, whose contribution to human history is so overwhelming that it can never be overemphasized. This species is *Saccharomyces cerevisiae*, commonly known, depending upon the circumstances, as baker's yeast or brewer's yeast. About 25 strains of *S. cerevisiae* are known, some of which have different physiological properties and were formerly recognized as separate species (Rainieri et al. 2003). In addition, there is a second species of yeast (*S. pasteurianus*) that is used in some types of brewing.

As briefly described in Chapter 3, yeasts have been used by humans for thousands of years to ferment alcoholic beverages and bake bread. It is impossible to imagine a world in which these two items did not exist, but this was once the case.

Bread

Archaeological evidence for baking dates back at least 9000 years, and some historians suggest that it might have been practiced several thousand years earlier. The very first breads produced were probably cooked versions of a grain-paste, made from ground cereal grains and water. Unleavened (yeast-free) breads of this type formed a staple in the diet of many early civilizations but would have been very different from most of the breads produced today. Just when the introduction of yeasts to the process of baking bread first took place is impossible to determine but may date back to prehistoric times. Yeasts occur everywhere, including on the surface of cereal grains, so any uncooked dough exposed to air for a period of time is likely to have picked up yeast and thus become leavened.

The function of yeast cells in the process of producing leavened bread is one of simply utilizing sugars present in the flour or added to the dough as an energy source and then giving off carbon dioxide and ethyl alcohol as a result. The carbon dioxide is trapped within tiny bubbles and results in the dough expanding (or rising). These bubbles are responsible for the texture of the bread, which is very different from that of unleavened bread.

The earliest evidence for leavened bread is from about 4000 BC in ancient Egypt. Using scanning electron microscopy, scientists have detected yeast cells in some remarkably preserved examples of ancient Egyptian bread. Intentionally adding yeasts to dough probably developed at a somewhat later date. It was certainly known to the Romans in the first century AD, since Pliny the Elder reported that the Gauls and Iberians used the foam skimmed from beer to produce "a lighter kind of bread than other peoples." It was more common, however, to retain a piece of the risen dough from the previous baking session to utilize as a "starter" for the next baking session. This particular technique is still practiced today.

Alcoholic Beverages

The use of yeasts for the production of fermented (or alcoholic) beverages probably began about 8000 years ago but could have occurred even earlier. The first alcoholic beverage developed was beer, and it is likely that the technique for brewing beer was independently advanced in various societies throughout the world.

The earliest known chemical evidence of beer dates to 3500 to 3100 BC in western Iran. The earliest Sumerian writings contain references to beer, and it is also mentioned in writings from ancient Egypt and Mesopotamia. Beer had been introduced to Europe by 3000 BC. Although appropriately called beer, the beverage produced in these ancient societies probably would not be recognized as such by most people today. For one thing, it did not contain hops, a relatively recent addition to beer brewing that was first referred to in Europe about 822 AD.

The first beer was brewed in essentially the same manner as that still used today. In brief, cereal grains are allowed to germinate and the resulting mash undergoes fermentation by yeast. Barley was the primary cereal grain used by early brewers, but wheat, corn, and rice also work for producing beer. Some historians believe that brewing beer provided a major incentive for the development of agriculture (and thus the growing of cereal grains) about 6000 BC. If this was the case, then it could be argued that both human civilization and the technological advances that the latter made possible might never have happened without fungi. The function of yeast cells in brewing is similar to what happens in baking except that ethyl alcohol is the more important product of fermentation.

Two other types of alcoholic beverages appeared later than beer. The first was mead, which is brewed from honey, and the second was wine, which is produced primarily from grape juice, although other juices can be used. It is not known for certain where wine was first produced, but solid evidence indicates that wine was being produced by 2300 BC in the Middle East. Wine quickly became a major item of trade, and this led to the development and use of special wine jars called amphorae (singular: amphora). These are often excavated from ancient shipwrecks, and much has been learned about wine and the wine trade in the ancient world.

Little is known about the origin of mead, but it is possible that this beverage predated the appearance of wine in portions of Europe. The first reference to mead dates from about 1700 BC, and the drink was common in both ancient Greece and Rome. Mead was the beverage of choice for early Germanic tribes in northern Europe, and the Vikings are reputed to have consumed a lot of mead before going into battle. It is possible, based on some of the information discussed in the context of *Amanita muscaria*, that the mead they consumed might have contained this fungus. If so, the mead would have represented a point of convergence for the two fungi that have played such prominent roles in human affairs.

12
Fossil Fungi

The Geologic Time Scale 228
Rhynie Chert Fossils 229
Chytrid Fossils 230
Ascomycete Fossils 231
Fossil Evidence of Mycorrhizae 232

Prototaxites 233
Fossils of Endomycorrhizae and Ectomycorrhizae 235
Amber-Preserved Fossils 236
Frozen Fossils 237

The fungi appear to have been around for a very long time. Mycologists John Taylor at the University of California and Mary Berbee at the University of British Columbia have attempted to date the evolutionary divergences of both the kingdom Fungi and the major assemblages of taxa within the kingdom by using the data available from molecular analyses and fossil evidence (Taylor and Berbee 2006). Their molecular analyses were based on what has become known as the "molecular clock" concept.

This concept, which has been widely used by biologists for several decades, is based upon the assumptions that (1) the rate at which changes occur in genes is constant over time and in different groups of organisms and (2) it is possible to relate the time at which two taxa diverged to the number of differences determined to exist in their DNA sequences. The divergence dates obtained in this manner are then compared with the dates known for relevant fossils (which serve as calibration points in time) of the groups of organisms being considered to come up with a final estimated date.

For the fungi, the estimated date of divergence is approximately 900 million years ago (mya). Until recently, the oldest known indisputable fossil evidence for fungi was from approximately 410 mya, but fossilized hyphalike structures are now known from sediments that possibly date back one billion years (Butterfield 2005). The latter date is supported by data from protein sequence analyses (Heckman et al. 2001).

The Geologic Time Scale

Just what do these dates mean in the context of the history of the earth? First of all, evidence from radiometric dating indicates that the earth itself is about 4.55 billion years old. The time that has passed since this event has been organized into a number of units on the geologic time scale. This scale was developed by geologists and other earth scientists. Each unit represents a span of time initially delimited by major geological or paleontological events but now determined by radioisotopic dates that provide absolute calibrations to the earlier event-based dates. These spans of time are of varying lengths and are constantly being refined as our technology and understanding of the geology improves. For example, the geologic time scale that I learned in college differs in a number of minor respects (dates assigned to the beginning and endings of certain spans of time) from the one used today.

In brief, the units of the geologic time scale most useful in a discussion of fossil fungi are known as eras and periods (Ogg et al. 2008). The first era (Precambrian) began with the formation of the earth 4.55 billion years ago and ended about 542 mya. This era encompasses most of the first 4 billion years of earth history, including a lower boundary that remains undefined because of the absence of sedimentary or metamorphic rocks. The second era (Paleozoic) extends from 542 to about 251 mya and was followed by the Mesozoic (251 to 65.5 mya) and the Cenozoic (65.5 mya to the present day).

The origin of life on the earth is estimated to have occurred about 3.3 billion years ago during the Precambrian, and the first evidence of eukaryotic cells dates to about 1.5 billion years ago. If the fungi diverged as a separate group some 900,000 mya, this would have placed this event during the latter part of the Precambrian. It should be noted that geologists often use the terms "Upper," "Middle," and "Lower" to refer to the rock record of eras and periods, while the terms "Early," "Middle," and "Late" are used for the respective time intervals associated with this rock record. As such, fungi would have first appeared in rocks from the Upper Precambrian.

The Paleozoic era is made up of six periods. These are the Cambrian (542 to 488 mya), Ordovician (488 to 444 mya), Silurian (444 to 416 mya), Devonian (416 to 359 mya), Carboniferous (359 to 299 mya), and Permian (299 to 251 mya). The earth during the Lower Paleozoic would have been significantly different from what it is today. Most of what was happening in terms of life would have been confined to the oceans of the world, and terrestrial landscapes would have been essentially barren with the exception of algal crusts and associated bacterial communities. At the end of the Paleozoic, some familiar groups of plants (for example, ferns, cycads, and conifers) were already present, but the terrestrial vegetation was dominated by groups (for example, seed ferns) that are now extinct.

The stage was set for the Mesozoic era, which is divided into three periods. These are the Triassic (251 to 199 mya), Jurassic (199 to 145.5 mya), and Cretaceous (145.5 to 65.5 mya). During the early part of the Mesozoic various groups of gymnosperms were both abundant and diverse. A major change began to occur with the appearance of the angiosperms (flowering plants) in the Early Cretaceous. As the dominant plants on the earth today, the angiosperms are linked to fungi in many different ways, as has been described throughout this book.

Rhynie Chert Fossils

In 1917, Robert Kidston and William Lang, two British paleobotanists, described a new fossil plant (*Rhynia gwynne-vaughani*) that was collected near the village of Rhynie in Aberdeenshire in northern Scotland. Interestingly, fossils of the plant were first noticed in a stone wall, and when their botanical significance was realized, the stones making up the wall were traced back to their source. This source turned out to be one of the more important Paleozoic fossil localities ever discovered. The rock that makes up the fossil-containing layers at this locality consists of chert, a type of very fine-grained crystalline quartz. The fossils themselves appear to be derived from plants that occurred in a tropical or subtropical marshlike setting that was subject to periodic inundation by water.

Rhynia gwynne-vaughani, along with several other associated plants (for example, *Horneophyton lignieri* and *Asteroxylon mackiei*) that were discovered later, apparently formed a dense growth that was disrupted by flooding several times, resulting in a series of deposits that had a total thickness of up to 8 feet (2.4 meters). The plants were preserved when mineral sediments settled around and on top of what was mostly organic plant material and compressed it. Over time, fossilization occurred through the replacement of the organic plant material by silica. The plant fossils are so superbly well preserved that anatomical features can be discerned for individual plant cells.

These fossils, from what has become known as the Rhynie chert, are particularly important because they date from about 410 mya in the Early Devonian, when vascular plants were still in the process of making the transition from water to land. Although fossil vascular plants were the first organisms reported, the Rhynie chert has also yielded fossils of algae, invertebrate animals, and—of primary concern for the information presented in this chapter—fungi.

Detailed studies of the fossil vascular plants from the Rhynie chert have revealed that representatives of at least four of the major phyla of fungi described earlier in this book had already appeared by 410 mya. These are the Chytridiomycota, Blastocladiomycota, Glomeromycota, and Ascomycota. Clearly, fungi had successfully invaded the land and begun to diversify by the Early Devonian. Moreover, the circumstances under which some of the fossils were found provide unequivocal evidence of fungus-plant interactions at this very early date.

Chytrid Fossils

Chytrids are the most common fossil fungi in the Rhynie chert, which might seem surprising because most chytrids are microscopic (Taylor et al. 1992). Nevertheless, chytrid fossils described from the Rhynie chert include an endobiotic form that is morphologically similar to members of the modern genus *Olpidium* and epibiotic forms that resemble some of the more common and widely distributed chytrids (for example, members of the genus *Rhizophydium*) one would expect to find in similar situations today (PLATE 118). For example, the fossils of the epibiotic form in the Rhynie chert are associated with the spores of a primitive vascular plant, which is rather similar to the colonization of pollen grains that occurs when the latter are used as "chytrid baits" in a laboratory exercise, as I have done on numerous occasions when teaching classes in plant diversity or mycology.

One of the chytrids from the Rhynie chert belongs to the Blastocladiomycota. This chytrid, which has been given the name of *Palaeoblastocladis milleri* (Remy et al. 1994), produces little tufts of hyphae that arise from the stomata or the cuticle and epidermis of stems of *Aglaophyton major*, one of the primitive plants described from the Rhynie chert (PLATE 119). The hyphae are nonseptate and are differentiated into two structural types. Moreover, both zoosporangia and gametangia can be identified.

All of the chytrids associated with fossil plants from the Rhynie chert appear to have been aquatic and undoubtedly occurred as saprotrophs on decaying plant material. Interestingly, these fossil chytrids are not the earliest records known for the group. Some fossil chytrid-like forms have been reported from the Late Precambrian of northern Russia.

Ascomycete Fossils

The perithecia of a fossil member of the Ascomycota have been found just below the epidermis of stems of *Asteroxylon mackiei*, one of the primitive vascular plants known from the Rhynie chert (Taylor et al. 2005). Each globose perithecium of this fungus, which has been given the name *Paleopyrenomycites devonicus*, is approximately 5/32 inch (400 µm) in diameter, and most appear to have developed just beneath the stomata. In mature examples, the wall of the perithecium consists of two distinct layers, and the inner layer gives rise to numerous, tightly packed, elongated asci containing between 16 and 32 ascospores. Also present are paraphyses, both interspersed among the asci and lining the inner layer of the neck of the perithecium.

The entire structure of some perithecia is so remarkably well preserved that microscopic features can be observed. For example, the ascus wall is thin and appears to consist of a single layer, and some asci possess a slight invagination that encircles the tip, suggesting the presence of some structural modification for spore release.

Perithecia similar in most respects to those produced by this ancient fungus, which is thought to have been a saprotroph, are commonly found in many modern pyrenomycetes that occur as saprotrophs or parasites on vascular plants. The most surprising aspect of this very early representative of this large and diverse group of fungi is that it does not appear to be especially primitive. This suggests that the ascomycetes had been around for some time.

Fossil Evidence of Mycorrhizae

As described in Chapter 9, the most widespread and important relationship that exists between fungi and plants in nature involves the establishment of what are known as mycorrhizal associations (or mycorrhizae). Because they are almost universal in plants, one might suspect that mycorrhizae may have evolved at a very early date, and there is evidence of this in some of the fossil plants from the Rhynie chert. For example, when very thin sections of the outer cortex of fossils of *Aglaophyton major* are examined, it is quite apparent that fungal hyphae are present. Some of these hyphae occupy the intracellular spaces in the cortex, but others clearly penetrate individual cells. In some of these cells, there are nicely preserved arbuscules that are essentially identical to those formed by modern vesicular-arbuscular mycorrhizal fungi.

It is clear that the Glomeromycota occurred some 410 mya, and recent evidence indicates that the phylum might date back as much as 60 million years earlier (Redecker et al. 2000). Fossil-bearing rocks collected from an ordinary roadcut in south-central Wisconsin also have been shown to contain fossilized fungal hyphae and spores, and the latter appear to be very similar to those produced by members of the modern genus *Glomus* (PLATE 120).

The sediments from which the rocks were formed have been dated to about 455 to 460 mya, which corresponds to the Late Ordovician. The fossil spores were not associated with plant fossils, as was the case in the Rhynie chert, but it is difficult to imagine that the spores could be from anything else other than a member of the Glomeromycota. If this is the case, fungi capable of forming mycorrhizae were quite possibly already present even before the first vascular plants appeared on the earth. Presumably, such fungi could have formed mycorrhizae with some of the first bryophyte-like plants to invade the land prior to the initial colonization of the terrestrial landscape by vascular plants.

Although the fossils mentioned above suggest that colonization of the land by plants may have started as early as the Late Ordovician, there is as yet very little fossil evidence to support such a hypothesis. These first bryophyte-like plants were undoubtedly rather small and had soft quickly decaying tissues of the type found in some modern bryophytes such as liverworts. Making the transition from living in water to surviving on land would have been difficult. In the terrestrial environment, water is often limited and sometimes absent, solar radiation (including potentially harmful ultraviolet rays) is much more intense, greater fluctuations in temperature occur, and obtaining required mineral nutrients is more difficult (Selosse and Le Tacon 1998).

In harsh environments that exist today, the mycorrhizal associations between fungi and plants often allow the latter to survive. It is certainly possible that a very early example of this type of association may have played a critical role in the success of the very first land plants.

Prototaxites

In the summer of 1843, during the course of a survey for coal and other mineral resources, the geologist William Logan collected a series of plant fossils from Lower Devonian strata exposed along the shores of Gaspé Bay in eastern Canada. Among the fossils was what appeared to be, on the basis of its size and general aspect, fossilized wood. Canadian paleobotanist Charles Dawson, who described the fossil, thought it was a fragment of a small tree, possibly a very early conifer. As a result, he chose the name *Prototaxites*, which literally means "early yew."

Additional fossils of *Prototaxites* from other localities throughout the world have established the fact that this organism was present from about 420 mya until approximately 359 mya, a period of time that encompasses the entire Devonian (Hueber 2001). Some of the fossils also indicated that *Prototaxites* was extraordinarily large for the time period during which it occurred (PLATE 121). Dawson, who visited the locality where Logan had collected the first specimen of *Prototaxites*, recorded in his notes that he had collected fragments of two large specimens, one 3 feet (0.9 m) and the other 2 feet (0.6 m) in diameter. This is larger than many modern trees. Based on measurements taken from other specimens, it is now known that the main axis of *Prototaxites* could reach a height of more than 20 feet (6.0 m).

In contrast, early vascular plants that lived at the same time and are so well preserved in the Rhynie chert were generally much smaller. For example, *Rhynia gwynne-vaughani* was only about 8 inches (20 cm) tall and its plant body had little in the way of supporting tissue. As such, for the greater part of its existence, *Prototaxites* was the largest and tallest terrestrial organism. In the Early and Middle Devonian it would have towered over everything else (PLATE 122), but by the Late Devonian *Prototaxites* would have been comparable in height to the earliest known trees (Stein et al. 2007). The mere size of this fossil fungus was enough to confound paleobotanists.

When Dawson examined thin sections made from well-preserved specimens of *Prototaxites* under a microscope, they showed what appeared to be a filamentous structure with numerous small tubes, which is not at all like what one would find in the tissue of a vascular plant. Initially, Dawson attributed these to the presence of a fungus, and the statement "Like mycelium of fungus" is recorded in his notes. As such, it was possible to interpret the fossil itself as somewhat decayed wood-like material. Other paleobotanists, upon examining specimens of *Prototaxites*, came to the conclusion that it was a giant alga, but it was difficult to imagine a terrestrial alga that could be so huge.

The taxonomic position of *Prototaxites* remained problematic for nearly a century and a half until a team of paleobotanists, including Kevin Boyce of the University of Chicago and Francis Hueber of the Department of Paleobiology at the National Museum of Natural History, used a technique called isotopic analysis to compare ratios of the two naturally occurring forms of carbon (carbon-12 and carbon-13) in *Prototaxites* and in other plants that lived in the same environment 410 mya (Boyce et al. 2007). For plants that photosynthesize, the ratios of the two forms of carbon in their tissues are essentially the same as the ratios that exist in the atmosphere, but this is not the case for non-photosynthesizing organisms. What Boyce and Hueber found was that *Prototaxites* did not photosynthesize. Since the "filaments" of which it was composed are almost certainly fungal hyphae, the conclusion that this organism was a very large ancient fungus is inescapable.

Just why *Prototaxites* grew so large remains an unanswered question, as does the nature of the structures for which there are fossils. Careful examination of thin sections from various regions of the main axis (probably best considered as a sporophore) of *Prototaxites* has revealed that there were three types of hyphae present. These included thick-walled, large, straight or flexuous skeletal hyphae that lacked septa and thin-walled, often extensively branched septate hyphae. The presence of several different types of hyphae is a common feature of many higher fungi, including some members of the Ascomycota, the phylum to which *Prototaxites* has been provisionally assigned by some paleobiologists. The sporophore of this ancient fungus was capable of perennial growth, and growth increments are apparent in thin sections taken of fossils of the sporophore. The growth increments are highly variable, and it is not known whether they reflect an annual increase or were produced whenever there was an extended period of unusually favorable conditions.

With such a massive sporophore, the mycelium of *Prototaxites* must have been extensive. Terrestrial plants living in the same environment would have produced organic material that could be decomposed, and it is possible that there were assemblages of bacteria and algae present that could have represented yet another energy source. The earth would have been a vastly different place 410 mya, so it is difficult to even speculate on the ecology of this unquestionably rather bizarre yet intriguing fungus.

Fossils of Endomycorrhizae and Ectomycorrhizae

Any effort to reconstruct the evolutionary history of fungi from the fossil record is limited to the evidence that is available. Fungi are not uncommon as fossils, but these fossils have not received the attention that has been devoted to other groups of organisms. A new fossil of a dinosaur is far more likely to be mentioned in the public media than a fossil of a fungus, even if the latter is more significant from an evolutionary standpoint.

One other problem with fungi is that most fossils are microscopic. For example, although microscopic fossil fungi are rather abundant and diverse in coal balls, which are lumps of petrified plant matter that occur in coal seams of Late Carboniferous (299 to 318 mya) age, they are often difficult to identify. Very few large fruiting bodies such as mushrooms have ever been found, with the few examples preserved in amber (to be discussed later in this chapter) representing noteworthy exceptions.

Nevertheless, there is enough evidence to date the first appearance of certain fungi in the fossil record. The oldest fossil lichen is from the Rhynie chert (Taylor et al. 1997), which means that the association of fungi and terrestrial algae is about as ancient as the association of fungi and land plants. The early associations of fungi and land plants were examples of what today we would recognize as endomycorrhizae.

In contrast, the oldest known fossil evidence of ectomycorrhizae is much more recent and dates only from the Middle Eocene, about 50 mya (LePage et al. 1997). It is suspected that these fungi appeared much earlier, possibly up to 130 mya during the Early Cretaceous. Only four fossil genera of agarics are currently accepted, with the oldest of these dating back to the Middle Cretaceous, about 90 mya (Hibbett et al. 2003). Interestingly, such fossilized fungi are remarkably like those produced by modern fungi, which certainly suggests that these fungi have had a long and stable existence.

Amber-Preserved Fossils

The science fiction film *Jurassic Park*, which came out in 1993, featured an amusement park of dinosaurs that had been created by cloning the genetic material found in mosquitoes that fed on dinosaur blood and then were preserved in amber. Although the science involved in the use of amber was unrealistic, the preservation aspects were not.

Amber is fossilized tree resin, the semi-liquid, often sticky substance produced by such plants as modern conifers. Almost all of the world's amber is 20 to 120 million years old. Because it is soft when first exuded by a tree, resin can sometimes "capture" plant parts (for example, leaves) or small organisms (including insects) that happen to get too close. Often, these can become completely enclosed in the resin, which impregnates their tissues before fossilization occurs. When pieces of amber are discovered as fossils, the structure or animal can be extraordinarily well preserved.

Amazingly enough, on very rare occasions, the fruiting bodies of macrofungi can be found in amber (Hibbett et al. 1997, 2003). Four such fossils are known, each of which has been described as a species new to science. Three of these (*Aureofungus yangiguaensis*, *Coprinites dominicana*, and *Protomycena electra* [PLATE 123]) are from Dominican Republic amber and are 15 to 20 million years old. The fourth (*Archaeomarasmius leggeti*) is from Atlantic Coastal Plain amber and is approximately 90 to 94 million years old. This date would place it back in the Middle Cretaceous long before the demise of the dinosaurs. As such, when this fruiting body was produced, it would have been part of the world of dinosaurs.

These four fossil macrofungi, all of which were assigned to the Basidiomycota, are not the only fungi and funguslike organisms that have been preserved in amber. Two of the more surprising finds have been nematode-trapping fungi and the fruiting body of a myxomycete. In the first of these, both the fungi themselves and their prey (several small nematodes) had been preserved in a piece of amber from southwestern France (Schmidt et al. 2007). The fossil deposit from which the amber was obtained is estimated to have been formed 100 mya during the Early Cretaceous. The fungi present were clearly nematode-trapping fungi, since some examples of the hyphal rings these organisms use to capture their prey can be identified. This fossil is significant because it indicates that the fungus-nematode association described in some detail in Chapter 10 has been in place for a very long time.

The fruiting body of a myxomycete turned up in a piece of Baltic amber that dates back to the Eocene, 35 to 50 mya (Dörfelt et al. 2003). Although only a single fruiting body is present, there is little question that the myxomycete involved is a member of the genus *Arcyria*. Indeed, the fossil appears to be remarkably similar in appearance to *A. denudata*, one of the more common and widespread myxomycetes in temperate forests today. This suggests that the myxomycetes that one would have encountered 35 mya might not have been very different from modern forms. The *Arcyria* in amber, which was described as a new species (*A. sulcata*), is the oldest known fossil myxomycete (PLATE 124). In fact, fossils of this group are exceedingly rare, and a specimen of *Stemonitis* (also from Baltic amber) appears to be the only other indisputable record of an ancient myxomycete (Domke 1952).

Frozen Fossils

As mentioned in Chapter 1, it has long been known that the spores of fungi, along with bacteria, viruses, various protists, and other small particles can be carried long distances by wind. Eventually, these particles fall to the earth, sometimes affixed to raindrops or snowflakes but often simply as dry deposition. If this happens in the polar regions, the particles have a chance of becoming incorporated within ice. This is especially true for the Antarctic and Greenland, where there are vast ice sheets. Once in the ice, the particles may remain frozen for a few years to millions of years (Castello and Rogers 2005).

As a result of their exposure to the intense cold, many fungal spores die, but others remain dormant. Fungal spores often have relatively thick walls, and these serve to protect the contents of the spore when subjected to stressful environmental conditions, even when the latter persist for a considerable period of time.

Viable fungal spores have been isolated from Antarctic ice cores more than 1.5 million years old and from Greenland ice cores 140,000 years old. In the study (Ma et al. 2005) that reported these results, ancient ice from the cores was allowed to melt, and aliquots of the meltwater used to inoculate culture plates prepared with media known to support the growth of fungi. The entire procedure was carried out under carefully controlled conditions to avoid contamination from present-day sources. After a period of incubation, living fungi were found to be present in a number of the plates. Several isolates of the yeast *Rhodotorula* were recovered from the very oldest ice core, which was 1.5 to 2.0 million years old. Only some of the spores present in the ice were found to be viable. Moreover, even when still viable, not all fungal spores will germinate in laboratory culture regardless of the types of media used.

Nevertheless, the fact remains that viable spores produced by fungi many thousands to millions of years ago are potentially capable of being released when the ice in which they have been entombed eventually melts. Because of their age, these viable spores would seem to qualify as fossils, but since they are still living, the spores certainly give a new and somewhat more literal meaning to the term "living fossil."

Glossary

aero-aquatic fungi group of hyphomycetes that colonize dead organic matter introduced to still and stagnant water; often characterized by the production of distinctive floating asexual propagules

aethalium (plural: **aethalia**) relatively large fruiting body produced by some myxomycetes

agaric fruiting body produced by some basidiomycetes; the hymenium is located on a series of platelike structures called gills

algal layer layer in the thallus of a lichen where the cells of the photobiont are located

ambrosia beetle a wood-boring beetle that cultivates fungi in the tunnels that it excavates in dead trees

anamorph asexual or imperfect stage of a fungus

annulus (plural: **annuli**) ring found on the stalk of some agarics and boletes

apothecium (plural: **apothecia**) cup- or disk-shaped fruiting body produced by some members of the phylum Ascomycota

aquatic hyphomycetes group of fungi that colonize dead organic matter introduced to flowing water systems; often characterized by the production of tetraradiate spores

ascocarp fruiting body produced by members of the Ascomycota

ascogenous hypha specialized hypha that gives rise to one or more asci

ascomycete informal term for members of the phylum Ascomycota

Ascomycota largest phylum in the kingdom Fungi

ascospore sexual spore produced by members of the Ascomycota

ascus (plural: **asci**) saclike structure within which ascospores (usually eight) are produced

asexual reproduction form of reproduction that does not involve genetic recombination

basidiocarp fruiting body produced by members of the Basidiomycota

basidiolichen lichen in which the mycobiont is a basidiomycete

basidiomycete informal term for members of the phylum Basidiomycota

Basidiomycota phylum in the kingdom Fungi to which the majority of the taxa that produce macroscopic fruiting bodies belong

basidiospore sexual spore produced by members of the Basidiomycota

basidium (plural: **basidia**) specialized hypha on which basidiospores (usually four) are produced

binomial taxonomic name for one kind of organism; it consists of two parts, with the first designating the genus and the second the species epithet

bioluminescent producing light through a biological process; glowing in the dark

budding type of asexual reproduction found in yeasts

capillitium (plural: **capillitia**) sterile, threadlike elements in the spore mass of many myxomycetes

chitin main component of the hyphal wall in the true fungi

chlamydospore thick-walled resting spore formed on vegetative hyphae

cleistothecium (plural: **cleistothecia**) type of fruiting body in which the asci are completely enclosed; characteristic of some members of the phylum Ascomycota

conidiophore specialized hypha upon which conidia are produced

conidium (plural: **conidia**) asexual spore produced by many ascomycetes and some basidiomycetes

coprophilous occurring on dung

cortex outermost layer (both upper and lower surfaces) of the thallus of a lichen

cortina partial veil that is weblike

crustose lichen thallus that is closely adherent to the surface of the substrate upon which it occurs

cystidium (plural: **cystidia**) sterile hypha occurring in the hymenium of some agarics

deliquescent to break down into a liquid; characteristic of many of the fungi traditionally assigned to the genus *Coprinus*

dikaryon two nuclei, each derived from a different source, in a single cell-like compartment of a hypha

diploid having two sets of chromosomes

ectomycorrhiza (plural: **ectomycorrhizae**) association between a plant root and a fungus in which the latter forms a sheath around the root

endomycorrhiza (plural: **endomycorrhizae**) association between a plant root and a fungus in which the hyphae of the fungus invade cells of the root

exoenzymes digestive enzymes released in the immediate environment of a fungal hypha

fairy ring ring of fruiting bodies formed around the outer perimeter of a mycelium

fleshy soft to the touch

foliose lichen thallus that consists of several distinct layers and is leaflike

fruiting body (or **fruit body**) general term for the specialized spore-producing structure produced by the mycelium of a fungus

fruticose lichen thallus that is upright or pendent; with a clearly three-dimensional structure

genus (plural: **genera**) a taxonomic group below family and above species; first of the two words in a binomial

gleba spore mass of a gasteromycete such as a puffball

haploid having one set of chromosomes

haustorium (plural: **haustoria**) specialized hypha produced by parasitic fungi; involved in the absorption of materials from the host

holomorph the whole fungus, including both asexual and sexual stages

host living organism with which a parasite or pathogen is associated

hymenium (plural: **hymenia**) fertile, spore-producing layer of a fruiting body

hypha (plural: **hyphae**) threadlike structure that makes up the body of a fungus

hypogeous fruiting body that occurs below the surface of the ground

lichen composite organism that consists of a fungus and an alga and/or cyanobacterium

medulla mass of hyphae that makes up most of the interior of the thallus of a lichen

meiosis type of cell division involved in sexual reproduction; the cells produced have half the number of chromosomes as the original cell

mitosis type of cell division in which the chromosomes duplicate and two daughter cells are produced, each with the same number of chromosomes as the original cell

mutualism type of association in which both of the organisms involved benefit

mycelium (plural: **mycelia**) collective name for the hyphae that make up the body of a fungus

mycobiont fungal component of a lichen

myxomycete a member of the largest and best-known group of slime molds

paraphysis (plural: **paraphyses**) sterile hypha found within the hymenium of the perithecium or apothecium of an ascomycete

parasite organism that meets its energy needs at the expense of another organism, referred to as the host

partial veil layer of tissue that extends from the edge of the pileus to the stalk in the fruiting bodies of some basidiomycetes; a covering over the developing hymenium

pathogen organism that interferes with the growth and development of another organism, causing a disease; a pathogen may result in the death of the organism it attacks

peridium (plural: **peridia**) outer covering of a fruiting body

perithecium (plural: **perithecia**) flask-shaped fruiting body produced by some members of the Ascomycota

phialide structure on a conidiophore that gives rise to a succession of conidia

photobiont photosynthesizing component of a lichen; an alga or cyanobacterium

pileus (plural: **pilei**) cap; part of the fruiting body held aloft by the stalk

plasmodiocarp branched, ring-shaped or netlike fruiting body produced by some members of the Myxomycota

plasmodium (plural: **plasmodia**) amoeboid, feeding stage of a myxomycete

pseudoaethalium (plural: **pseudoaethalia**) fruiting body that consists of a mass of sporangia tightly packed together; produced by members of the Myxomycota

pseudothecium (plural: **pseudothecia**) type of perithecium-like fruiting body produced by the loculoascomycetes; they resemble perithecia but have a different type of development

resupinate occurring flat on the surface of the substrate

rhizoid rootlike extension of a hypha

rhizomorph thick rootlike strand of hyphae

saprotroph organism that meets its energy needs from dead organic matter

septate containing or divided by cross walls or septa

septum (plural: **septa**) cross wall or partition in a hypha

sexual reproduction form of reproduction that involves genetic recombination

species a particular kind of fungus, designated by a binomial

specific epithet second of the two words in a binomial

sporangiophore hypha upon which a sporangium occurs

sporangium (plural: **sporangia**) relatively simple reproductive structure produced by some fungi and myxomycetes

spore reproductive cell produced by fungi and slime molds

squamulose lichen in which the thallus appears to consist of numerous small scalelike structures

stalk portion of a fruiting body that holds up the cap or sporotheca

sterigma (plural: **sterigmata**) small extension of a basidium to which a basidiospore is attached

stroma (plural: **stromata**) compact mass of hyphae on or within which individual fruiting structures occur in some ascomycetes

synnema (plural: **synnemata**) group of tightly packed conidiophores that form a single structure

teleomorph sexual or perfect stage of a fungus

teliospore thick-walled spore produced by rusts and smuts

telium (plural: **telia**) structure within which teliospores are produced

thallus (plural: **thalli**) body of a lichen

universal veil layer of tissue that surrounds the developing fruiting body in some basidiomycetes

VAM fungi vesicular-arbuscular mycorrhizal fungi

volva cuplike structure left at the base of the stalk when the universal veil ruptures

yeast single-celled fungus that reproduces by budding

zoosporangium (plural: **zoosporangia**) structure in which zoospores are produced

zoospore motile, asexual spore produced by some aquatic fungi and funguslike organisms

zygote diploid cell resulting from the fusion of two haploid cells

References

Aaronson, S. 2000. Fungi. In *The Cambridge World History of Food*. Eds. K. F. Kiple and K. C. Ornelas. Cambridge, United Kingdom: Cambridge University Press. 313–336.

Alexander, H. M., P. H. Thrall, J. Antonovics, A. M. Jarosz, and P. V. Oudemans. 1996. Population dynamics and genetics of plant disease: a case study of anther-smut disease. *Ecology* 77: 990–996.

Alexopoulos, C. J., and C. W. Mims. 1979. *Introductory Mycology*. 3rd ed. New York: John Wiley & Sons.

Alexopoulos, C. J., C. W. Mims, and M. Blackwell. 1996. *Introductory Mycology*. 4th ed. New York: John Wiley & Sons.

Allegro, J. M. 1970. *The Sacred Mushroom and the Cross*. New York: Bantam Books.

Anagnostakis, S. L. 2001. The effect of multiple importations of pests and pathogens on a native tree. *Biological Invasions* 3: 245–254.

Antoine, M. E. 2004. An ecophysiological approach to quantifying nitrogen fixation by *Lobaria oregana*. *The Bryologist* 107: 82–87.

Arora, D. 1979. *Mushrooms Demystified*. Berkeley, California: Ten Speed Press.

Banno, I. 1967. Studies on the sexuality of *Rhodotorula*. *The Journal of General and Applied Microbiology* 13: 167–196.

Barnett, H. L., and B. B. Hunter. 1998. *Illustrated Genera of Imperfect Fungi*. 4th ed. St. Paul, Minnesota: The American Phytopathological Society Press.

Begerow, D., M. Stoll, and R. Bauer. 2006. A phylogenetic hypothesis of Ustilaginomycotina based on multiple gene analyses and morphological data. *Mycologia* 98: 906–916.

Bell, A. 2005. *An Illustrated Guide to the Coprophilous Ascomycetes of Australia*. VBS Biodiversity Series 3. Utrecht, The Netherlands: Centraalbureau voor Schimmelcultures.

Berkeley, M. J. 1857. *Introduction to Cryptogamic Botany*. London: H. Bailliere.

Bills, G. F., and J. D. Polishook. 1994. Abundance and diversity of microfungi in leaf litter of a lowland rain forest in Costa Rica. *Mycologia* 86: 187–198.

Binion, D. E., S. L. Stephenson, W. C. Roody, H. H. Burdsall, O. K. Miller, Jr., and L. N. Vasilyeva. 2008. *Macrofungi Associated with Oaks of Eastern North America*. Morgantown, West Virginia: West Virginia University Press.

Blackwell, M., D. S. Hibbett, J. W. Taylor, and J. W. Spatafora. 2006. Research Coordination Networks: a phylogeny for kingdom Fungi (Deep Hypha). *Mycologia* 98: 829–837.

Boddy, L., O. M. Gibbon, and M. A. Grundy. 1985. Ecology of *Daldinia concentrica*: effect of abiotic variables on mycelial extension and interspecific interactions. *Transactions of the British Mycological Society* 85: 201–211.

Boyce, C. K., C. L. Hotton, M. L. Fogel, G. D. Cody, R. M. Hazen, A. H. Knoll, and F. M. Hueber. 2007. Devonian landscape heterogeneity recorded by a giant fungus. *Geology* 35: 399–402.

Brodie, H. J. 1975. *The Bird's Nest Fungi*. Toronto: Toronto University Press.

Brodo, I. M., S. D. Sharnoff, and S. Sharnoff. 2001. *Lichens of North America*. New Haven, Connecticut: Yale University Press.

Bruns, T. D. 1984. Insect mycophagy in the Boletales: fungivore diversity and the mushroom habitat. In *Fungus-Insect Relationships: Perspectives in Ecology and Evolution*. Eds. Q. Wheeler and M. Blackwell. New York: Columbia University Press. 91–129.

Butterfield, N. J. 2005. Probable Proterozoic fungi. *Paleobiology* 31: 165–182.

Cahill, D. M., J. E. Rookes, B. A. Wilson, L. Gibson, and K. L. McDougall. 2008. *Phytophthora cinnamomi* and Australia's biodiversity: impacts, predictions and progress towards control. *Australian Journal of Botany* 56: 279–310.

Castello, J. D., and S. O. Rogers, eds. 2005. *Life in Ancient Ice*. Princeton, New Jersey: Princeton University Press.

Claridge, A. W., and T. W. May. 1994. Mycophagy among Australian mammals. *Australian Journal of Ecology* 19: 251–275.

de Bary, A. 1866. *Morphologie und Physiologie der Pilze, Flechten und Myxomyceten*. Leipzig, Germany: Wilhelm Engelmann.

Deacon, J. 2006. *Fungal Biology*. 4th ed. Oxford, United Kingdom: Blackwell Publishing.

Desjardin, D. E., M. Capelari, and C. Stevani. 2007. Bioluminescent *Mycena* species from São Paulo, Brazil. *Mycologia* 99: 317–331.

Dice, L. R. 1921. Notes on the mammals of interior Alaska. *Journal of Mammalogy* 2: 20–28.

Domke, W. 1952. Der erste sichere Fund eines Myxomyceten im Baltischen Bernstein (*Stemonitis slendens* Rost. Fa. Succini fa. *succini* fa. nov. foss.). *Mitteilungen aus dem Geologischen Staatsinstitut in Hamburg* 21: 154–161.

Dörfelt, H., A. R. Schmidt, P. Ullmann, and J. Wunderlich. 2003. The oldest fossil myxogastroid slime mould. *Mycological Research* 107: 123–126.

Dugan, F. M. 2008. *Fungi in the Ancient World: How Mushrooms, Mildews, Molds, and Yeast Shaped the Early Civilizations of Europe, the Mediterranean, and the Near East*. St. Paul, Minnesota: The American Phytopathological Society Press.

Ferguson, B. A., T. A. Dreisbach, C. G. Parks, G. N. Filip, and C. L. Schmitt. 2003. Coarse-scale population structure of pathogenic *Armillaria* species in a mixed-conifer forest in the Blue Mountains of northeast Oregon. *Canadian Journal of Forest Research* 33: 612–623.

Fossey, D. 1983. *Gorillas in the Mist*. Boston, Massachusetts: Houghton Mifflin.

Frank, A. B. 1885. Über die auf Würzelsymbiose beruhende Ehrnährung gewisser Bäum durch unterirdische Pilze. *Berichte der Deutschen Botanischen Gesellschaft* 3: 128–145.

Fröhlich, J., and K. D. Hyde. 1999. Biodiversity of palm fungi in the tropics: are global fungal diversity estimates realistic? *Biodiversity and Conservation* 8: 977–1004.

Gilbertson, R. L. 1984. Relationships between insects and wood-rotting Basidiomycetes. In *Fungus-Insect Relationships: Perspectives in Ecology and Evolution*. Eds. Q. Wheeler and M. Blackwell. New York: Columbia University Press. 130–165.

Griffin D. W., C. A. Kellogg, and E. A. Shinn. 2001. Dust in the wind: long-range transport of dust in the atmosphere and its implications for global public and ecosystem health. *Global Change and Human Health* 2: 20–33.

Hagiwara, H., and A. Someya. 1992. Killer activity observed in dictyostelid cellular slime molds. *Bulletin of the Natural Science Museum, Tokyo, Series B (Botany)* 18: 17–22.

Hale, M. E., Jr. 1983. *The Biology of Lichens.* 3rd ed. London: Edward Arnold.

Hall, I. R., G. T. Brown, and A. Zambonelli. 2007. *Taming the Truffle: The History, Lore and Science of the Ultimate Mushroom.* Portland, Oregon: Timber Press.

Hall, I., S. L. Stephenson, P. K. Buchanan, W. Yun, and A. L. J. Cole. 2003. *Edible and Poisonous Mushrooms of the World.* Portland, Oregon: Timber Press.

Hartig, T. 1840. *Vollständige Naturgeschichte der forstlichen Culturpflanzen Deutschlands.* Berlin, Germany: Förstner'sche Verlagsbuchhandlung.

Hawksworth, D. L. 2001. The magnitude of fungal diversity: the 1.5 million species estimate revisited. *Mycological Research* 105: 1422–1432.

Hawksworth, D. L., and F. Rose. 1970. Qualitative scale for estimating sulphur dioxide air pollution in England and Wales using epiphytic lichens. *Nature* 227: 145–148.

Heckman, D. S., D. M. Geiser, B. R. Eidell, R. L. Stauffer, N. L. Kardos, and S. B. Hedges. 2001. Molecular evidence for the early colonization of land by fungi and plants. *Science* 293: 1129–1133.

Hennebert, G. L., and L. K. Weresub. 1977. Terms for states and forms of Fungi, their names and types. *Mycotaxon* 6: 207–211.

Hibbett, D. S., M. Binder, Z. Wang, and Y. Goldman. 2003. Another fossil agaric from Dominican amber. *Mycologia* 95: 685–687.

Hibbett, D. S., D. Grimaldi, and M. J. Donogue. 1997. Fossil mushrooms from Cretaceous and Miocene ambers and the evolution of homobasidiomycetes. *American Journal of Botany* 84: 981–991.

Hibbett, D. S., and R. G. Thorn. 1994. Nematode-trapping in *Pleurotus tuberregium*. *Mycologia* 86: 696–699.

Hobbs, C. 1995. *Medical Mushrooms: An Exploration of Tradition, Healing, and Culture.* 3rd ed. Santa Cruz, California: Botanica Press.

Horak, E., and O. K. Miller. 1992. *Phaeogalea* and *Galerina* in arctic-subarctic Alaska (USA) and the Yukon Territory (Canada). *Canadian Journal of Botany* 70: 414–433.

Houghton, W. 1885. Notices of fungi in Greek and Latin authors. *Annals and Magazine of Natural History* 15: 22–49.

Hueber, F. M. 2001. Rotted wood—alga—fungus: the history and life of *Prototaxites* Dawson 1859. *Review of Palaeobotany and Palynology* 116: 123–158.

Ingold, C. T. 1942. Aquatic hyphomycetes of decaying alder leaves. *Transactions of the British Mycological Society* 25: 339–417.

Ingold, C. T., and H. J. Hudson. 1993. *The Biology of Fungi.* 6th ed. London: Chapman & Hall.

Kendrick, B. 2000. *The Fifth Kingdom.* 3rd ed. Newburyport, Massachusetts: Focus Publishing.

Kiffer, E., and M. Morelet. 2000. *The Deuteromycetes, Mitosporic Fungi: Classification and Generic Keys.* Enfield, New Hampshire: Science Publishers.

Landolt, J. C., S. L. Stephenson, and J. C. Cavender. 2006. Distribution and ecology of dictyostelid cellular slime molds in the Great Smoky Mountains National Park. *Mycologia* 98: 541–549.

Landolt, J. C., S. L. Stephenson, and M. E. Slay. 2006. Dictyostelid cellular slime molds from caves. *Cave and Karst Journal* 68: 22–26.

Laursen, G. A., R. D. Seppelt, and M. Hallam. 2003. Cycles in the forest: mammals, mushrooms, mycophagy, mycoses, and mycorrhizae. *Alaska Park Science* (Winter): 13–19.

Lawrey, J. D. 1983. Lichen herbivore preference: a test of two hypotheses. *American Journal of Botany* 70: 1188–1194.

LePage, B. A., R. S. Currah, R. A. Stockney, and G. W. Rothwell. 1997. Fossil ectomycorrhizae from the middle Eocene. *American Journal of Botany* 84: 410–412.

Lincoff, G. H. 1981. *The Audubon Society Field Guide to North American Mushrooms.* New York: Alfred A. Knopf.

Longcore, J. E., A. P. Pessier, and D. K. Nichols. 1999. *Batrachochytrium dendrobatidis* gen. et sp. nov., a chytrid pathogenic to amphibians. *Mycologia* 91: 219–227.

Ma, L.-J., C. M. Catranis, W. T. Starmer, and S. O. Rogers. 2005. The significance and implications of the discovery of filamentous fungi in glacial ice. In *Life in Ancient Ice.* Eds. J. D. Castello and S. O. Rogers. Princeton, New Jersey: Princeton University Press. 159–180.

Martin, G. W. 1927. Basidia and spores of the Nidulariaceae. *Mycologia* 19: 239–247.

Martin, G. W., and C. J. Alexopoulos. 1969. *The Myxomycetes.* Iowa City: University of Iowa Press.

Maser, C., A. W. Claridge, and J. M. Trappe. 2008. *Trees, Truffles, and Beasts.* New Brunswick, New Jersey: Rutgers University Press.

Maser, Z., C. Maser, and J. M. Trappe. 1985. Food habits of the northern flying squirrel (*Glaucomys sabrinus*) in Oregon. *Canadian Journal of Zoology* 63: 1085–1088.

McCune, B., and L. Geiser. 1997. *Macrolichens of the Pacific Northwest*. Corvallis, Oregon: Oregon State University Press.

Meier, F. C., and C. A. Lindbergh. 1935. Collecting micro-organisms from the Arctic atmosphere. *Science Monthly* 40: 5–20.

Miller, O. K., Jr. 1973. *Mushrooms of North America*. New York: E. P. Dutton.

Miller, O. K., Jr., and H. H. Miller. 2006. *North American Mushrooms: A Field Guide to Edible and Inedible Species*. Helena, Montana: FalconGuide.

Nagahama, T. 2006. Yeast biodiversity in freshwater, marine and deep-sea environments. In *Biodiversity and Ecophysiology of Yeasts*. Eds. C. Rosa and G. Péter. Heidelberg, Germany: Springer-Verlag. 241–262.

O'Brien, M., K. F. Nielsen, P. O'Kiely, P. D. Forristal, H. T. Fuller, and J. C. Frisvad. 2006. Mycotoxins and other secondary metabolites produced in vitro by *Penicillium paneum* Frisvad and *Penicillium roqueforti* Thom isolated from baled grass silage in Ireland. *Journal of Agricultural and Food Chemistry* 54: 9268–9276.

Ogg, J. G., G. Ogg, and F. M. Gradstein. 2008. *The Concise Geologic Time Scale*. New York: Cambridge University Press.

Olive, L. S. 1975. *The Mycetozoans*. New York: Academic Press.

Rainieri, S., C. Zambonelli, and Y. Kaneko. 2003. *Saccharomyces* sensu stricto: systematics, genetic diversity and evolution. *Journal of Bioscience and Bioengineering* 96: 1–9.

Raper, K. B. 1984. *The Dictyostelids*. Princeton, New Jersey: Princeton University Press.

Rawlins, J. E. 1984. Mycophagy in Lepidoptera. In *Fungus-Insect Relationships: Perspectives in Ecology and Evolution*. Eds. Q. Wheeler and M. Blackwell. New York: Columbia University Press. 382–423.

Read, D. J., J. G. Duckett, R. Francis, R. Ligrone, and A. Russell. 2000. Symbiotic fungal associations in "lower" land plants. *Philosophical Transactions of the Royal Society of London* 355: 815–831.

Redecker, D., R. Kodner, and L. E. Graham. 2000. Glomalean fungi from the Ordovician. *Science* 289: 1920–1921.

Remy, W., T. N. Taylor, and H. Hass. 1994. Early Devonian fungi: a blastocladalean fungus with sexual reproduction. *American Journal of Botany* 81: 690–702.

Rice, M., and D. Beebee. 1980. *Mushrooms for Color*. Eureka, California: Mad River Press.

Rogerson, C. T., and S. L. Stephenson. 1993. Myxomyceticolous fungi. *Mycologia* 85: 456–469.

Roody, W. C. 2003. *Mushrooms of West Virginia and the Central Appalachians*. Lexington, Kentucky: The University Press of Kentucky.

Saccardo, P. A. 1882–1931. *Sylloge fungorum omnium hucusque cognitorum*. 25 vols. Variously published in Padua, Paris, Padova, and Avellino. (Authorship of volumes varies.)

Schmidt, A. R., H. Dörfelt, and V. Perrichot. 2007. Carnivorous fungi from Cretaceous amber. *Science* 318: 1743.

Schwendener, S. 1867. Über die wahre Natur der Flechtengonidien. *Verhandlungen der Schweizerischen Naturforschenden Gesellschaft* 51: 88–90.

Selosse, M. A., and F. Le Tacon. 1998. The land flora: a phototroph-fungus partnership? *Trends in Ecology and Evolution* 13: 15–20.

Stamets, P. 1996. *Psilocybin Mushrooms of the World*. Berkeley, California: Ten Speed Press.

Stein, W. E., F. Mannolini, L. V. Hernick, E. Landing, and C. M. Berry. 2007. Giant cladoxylopsid trees resolve the enigma of the earth's earliest forest stumps at Gilboa. *Nature* 446: 904–907.

Stephenson, S. L., and J. C. Landolt. 1996. The vertical distribution of dictyostelids and myxomycetes in the soil/litter microhabitat. *Nova Hedwigia* 62: 105–117.

Stephenson, S. L., and J. C. Landolt. 1998. Dictyostelid cellular slime molds in canopy soils of tropical forests. *Biotropica* 30: 657–661.

Stephenson, S. L., and H. Stempen. 1994. *Myxomycetes: a Handbook of Slime Molds*. Portland, Oregon: Timber Press.

Suh, S.-O., J. V. McHugh, D. O. Pollock, and M. Blackwell. 2005. The beetle gut: a hyperdiverse source of novel yeasts. *Mycological Research* 109: 261–265.

Suh, S.-O., N. H. Nguyen, and M. Blackwell. 2008. Yeasts isolated from plant-associated beetles and other insects: seven novel *Candida* species near *Candida albicans*. *FEMS Yeast Research* 8: 88–102.

Tapper, R. 1981. Direct measurement of translocation of carbohydrate in the lichen, *Cladonia convoluta*, by quantitative autoradiography. *New Phytologist* 89: 429–437.

Taylor, J. W., and M. L. Berbee. 2006. Dating divergences in the Fungal Tree of Life: review and new analyses. *Mycologia* 98: 838–849.

Taylor, T. N., H. Hass, and H. Kerp. 1997. A cyanolichen from the Lower Devonian Rhynie chert. *American Journal of Botany* 84: 992–1004.

Taylor, T. N., H. Hass, H. Kerp, M. Krings, and R. T. Hanlin. 2005. Perithecial ascomycetes from the 400 million year old Rhynie chert: an example of ancestral polymorphism. *Mycologia* 97: 269–285.

Taylor, T. N., W. Remy, and H. Hass. 1992. Fungi from the Lower Devonian Rhynie chert: Chytridiomycetes. *American Journal of Botany* 79: 1233–1241.

Taylor, T. N., W. Remy, H. Hass, and H. Kerp. 1995. Fossil arbuscular mycorrhizae from the Early Devonian. *Mycologia* 87: 560–573.

Thaxter, R. 1896, 1908, 1924, 1926, 1931. Contribution towards a monograph of the Laboulbeniaceae. Parts 1–5. Memoirs of the American Academy of Arts at Sciences. 12: 187–429; 13: 217–469; 14: 309–426; 15: 427–580; 16: 1–435.

Thrall, P. H., A. Biere, and J. Antonovics. 1993. Plant life-history and disease susceptibility—the occurrence of *Ustilago violacea* on different species within the Caryophyllaceae. *Journal of Ecology* 81: 489–498.

Totter, A. 1972. *Sylloge fungorum omnium hucusque cognitorum*. Vol. 26. New York and London: Johnson Reprint Corporation. (A final volume of the *Sylloge fungorum omnium hucusque cognitorum* assembled and translated by Edith Cash.)

Trappe, J. M. 2005. A. B. Frank and mycorrhizae: the challenge to evolutionary and ecologic theory. *Mycorrhiza* 15: 277–281.

Tschermak-Woess, E. 2000. The algal partner. In *CRC Handbook of Lichenology*. Vol. 1. Ed. M. Galun. Boca Raston, Florida: CRC Press. 39–92.

Wasson, R. G. 1968. *Soma: Divine Mushroom of Immortality*. New York: Harcourt Brace Jovanovich.

Wasson, V. P., and R. G. Wasson. 1957. *Mushrooms, Russia and History*. Vols. 1 and 2. New York: Pantheon.

Webster, J., and R. W. S. Weber. 2007. *Introduction to Fungi*. 3rd ed. Cambridge, United Kingdom: Cambridge University Press.

Wen, J. 1999. Evaluation of eastern Asia and eastern North American disjunct distributions in flowering plants. *Annual Review of Ecology and Systematics* 30: 421–455.

Yang, S., and D. H. Pfister. 2006. *Monotropa uniflora* plants of eastern Massachusetts form mycorrhizae with a diversity of russulacean fungi. *Mycologia* 98: 535–540.

Index

Abert's squirrel, 204
Achlya, 34
Acytostelium, 158
Adelanthaceae, 177
aeciospore, 181
aecium, 181
aero-aquatic fungus, 42, 44; Plate 15
aethalium, 149, 150
aflatoxin, 59
Africa, 23, 31, 90, 189, 190
African clawed frog, see *Xenopus laevis*
agaric, 104–114, 118, 168, 187, 201, 208, 236
Agaricaceae, 19, 107, 108, 112
agaricales, 103–105, 114, 115, 120, 189
Agaricomycotina, 20, 102, 103
Agaricus, 19, 112, 113
  bisporus, 19, 20, 101, 217, 218
  bitorquis, 113
  blazei, 218
  campestris, 19, 210, 217; Plate 4
Agathidium, 153
Aglaophyton major, 231, 232; Plate 119
Alaska, 87, 114, 137, 141, 154, 158, 168, 203, 204
alder, see *Alnus*
alfalfa, 28
alga, 23, 26, 143, 148, 186, 207
  fossils of, 229, 230, 234, 235
  lichens and, 21, 128, 130–134, 136, 139, 140–141
  parasites of, 28, 33
  photosynthesis and, 138, 164
Alice's Adventures in Wonderland (Alice in Wonderland), 210–212
alkaloid, 72, 177
Allegro, John, 223
Allomyces, 32
Alnus, 99; Plate 94
Alps, 151, 207
Alternaria, Plate 5
Amanita, 108, 109, 171, 217
  bisporigera, 108; Plate 48
  caesarea, 108, 217
  jacksonii, 108
  muscaria, 108, 213, 223, 224, 226; Plate 117
  phalloides, 108
  rubescens, 108
Amanitaceae, 107, 108
ambrosia beetle, 190–193, 196; Plate 108
Ambrosiella, 192
American ash, see *Fraxinus americana*
American beech, see *Fagus grandifolia*
American Caesar's mushroom, see *Amanita caesarea*
American chestnut, see *Castanea dentata*
American Chestnut Foundation, 214

American elm, see *Ulmus americana*
Amoebidiales, 40
amphibian, 29–31, 113, 159
Amylostereum, 193
anamorph, 54–58, 60, 63, 74–76, 78–80, 125, 154, 192, 196
angiosperm, 164, 179, 229
anise-scented clitocybe, see *Clitocybe odora*
Anisotoma, 153
Anna's hummingbird, see *Calypte anna*
annulus, 106, 108, 109
Antarctic, 64, 129, 143, 237, 238
antheridium, 35, 36
anther-smut, see *Microbotryum dianthorum*
aphid, 53, 71
aphyllophorales, 103, 104, 115, 120
Apiosporina morbosa, 82; Plate 33
apothecium, 69, 75, 85, 86, 93, 95, 140; Plate 42
Appalachian Mountains, 86, 92, 93, 158, 184
appendage, 41, 43, 75, 84
apple, see *Pyrus malus*
apple orchard, 86
apple scab, see *Venturia inaequalis*
appressorium, 131
arbuscule, 172, 174, 232
Archaeomarasmius leggeti, 236
Arctic tumbleweed, see *Masonhalea richardsonii*
Arcyria, 152, 154, 237
  denudata, 237
  sulcata, 237; Plate 124
Aristotle, 208
Arkansas, 158
Armillaria, 110
  mellea, 110, 207, 208
  ostoyae, 15
Arnelliaceae, 177
Arrhenatherum elatius, 184

Arrhenia retiruga, 168
Arthrobotrys oligospora, 198; Plate 109
artist's conk, see *Ganoderma applanatum*
Ascobolaceae, 85, 94
Ascobolus, 94, 169; Plate 42
  calesco, Plate 22
ascocarp, 56, 67
ascogenous hypha, 67, 68, 101
ascoma, 67
ascomycete(s), 78, 83, 130, 192, 196
  as decomposers, 164, 167, 169
  asexual reproduction in, 53–56
  classification of, 20
  fossils of, 231
  mitosporic members, 42, 60–61
  morphology of, 68–70, 73
  mycorrhizal associations of, 171, 174, 177
Ascomycota, 42–45, 46, 61–65, 85, 101–102, 169
  classification of, 20, 56–57, 104
  fossils of, 230, 231
  morphology of, 22, 53, 234
  reproduction in, 53–56, 67–69
ascospore, 56, 92, 99, 140, 231; Plate 22
  ascus and, 53, 54
  development of, 67–68, 79
  discharge of, 62, 75, 84, 89, 95, 96, 97
  germination of, 102
  number of, 98, 101
  role in diseases, 80, 82–83
ascus, 75, 79, 82, 89, 95, 98; Plate 22
  apical ring of, 96, 97
  Ascomycota and, 53–54, 56, 101
  development of, 62, 67–69, 83, 84
  disintegration of, 80, 140
  fossils of, 231
  operculate type, 76
Asellariales, 41

Aseroë rubra, 125; Plate 2
asexual reproduction, 17, 62, 140, 156
Asia, 23, 24, 73, 81, 99, 126, 174, 182, 189, 213, 217–220
aspen, see *Populus tremuloides*
aspen scaber stalk, see *Leccinum insigne*
aspergillosis, 60
Aspergillus, 58–60, 75; Plate 23
  flavus, 59
  fumigatus, 60
  niger, 60
  oryzae, 60
Aspicilia esculenta, 137
Asteroxylon mackiei, 229, 231
Atelopus varius, 30
Atlantic Ocean, 23
Atta, 187
attines, see *Atta*
Aureofungus yangiguaensis, 236
Auricularia, 126, 218
  auricula-judae, 126
  polytricha, 126
Australia, 23, 30, 31, 40, 73, 108, 125, 141, 151, 170, 202, 203, 205, 219
Australian Antarctic Division, 94
avec la mouche, 89
avocado, 40
azalea, 71, 184
Aztecs, 221

bacteria, 164, 177, 178, 188, 207, 220
  air-borne, 23, 237
  as a food source, 148, 155–156, 158, 161, 196
  distribution in soil, 161–162
  in Paleozoic era, 229, 235
badger, 89
Baeospora myosura, 168

baker's yeast, see *Saccharomyces cerevisiae*
Baltic amber, 237; Plate 124
bamboo mushroom, see *Dictyophora indusiata*
barberry, see *Berberis vulgaris*
Barnett, Harold, 61
Baroque style of art, 208
basidiomycete(s), 20, 127, 164, 193
  classification of, 125
  dung-loving, 169
  lichens belonging to, 130
  mycorrhizal associations of, 171
  nematode-trapping, 196
  reproduction in, 101–102
  wood-decaying, 199
Basidiomycota, 77, 101, 177, 178
  classification of, 20–21, 102
  dikaryotic condition in, 68, 101
  dung-loving members of, 169, 170
  fossils of, 236
  fruiting bodies of, 88
  members of, 22, 47
  relationship with Ascomycota, 54, 55, 61, 67
  yeasts belonging to, 63–65
basidiospore, 104, 114, 115, 179–181, 208; Plate 47
  development of, 101, 106
  discharge of, 103, 105, 120
  germination of, 102
  numbers of, 119
basidium, 101, 103–105, 119, 120, 125, 180; Plate 47
bat, 159
Batrachochytrium dendrobatidis, 30, 31; Plate 9
bear's head tooth fungus, see *Hericium americanum*
bearded tooth, see *Hericium erinaceus*
beech, see *Fagus grandifolia*

beech bark disease, 78, 81
beech strawberry, see *Cyttaria gunnii*
beer, 225, 226
Bell, Ann, 170
Berbee, Mary, 227
Berberis vulgaris, 179
Berkeley, Miles Joseph, 23
Bettongia gaimardi, 203
Bettongia penicillata, 203
Bettongia tropica, 203
Betulaceae, 98
biflagellate, 33
Bills, Gerald, 57
binomial, 19
bioluminescence, 207, 208
birch, 98, 166, 171
bird's nest fungus, see *Cyathus striatus*
Biscogniauxia, 97
Bisporella citrina, 76
bitunicate, 82
black bear, 214
black knot of cherry, see *Apiosporina morbosa*
black morel, see *Morchella elata*
Blackwell, Meredith, 194
Blastocladiella, 33
  emersonii, 33˙
Blastocladiomycota, 20, 27, 31–33, 230
bleeding mycena, see *Mycena haematopus*
blue poison dart frog, see *Dendrobates azureus*
blueberry, 184
blue-strain of wood, see *Chlorociboria aeruginascens*
blusher, see *Amanita rubescens*
Boletaceae, 114
Boletales, 114
bolete, 22, 104, 113–115, 199–201
Boletus, 114, 171
  edulis, 114, 217; Plate 55

bonnet mold, see *Spinellus fusiger*
Bordeaux mixture, 37
Botrytis, 76
  cinerea, 76
Boyce, Kevin, 234
bracket fungus, 116, 207, 217
bread, 72, 137
  mold on, 22, 47, 48, 58, 220
  yeast in, 61, 62, 224–225
Brefeld, Oskar, 155
Brefeldia maxima, 149
brewer's yeast, see *Saccharomyces cerevisiae*
bridal veil fungus, see *Dictyophora multicolor*
bristlecone pine, see *Pinus longaeva*
British soldier lichen, 140; Plate 79
Brodie, Harold, 123
brood chamber, 192
Brown, Gordon, 92
brown beard lichen, see *Bryoria*
brown false morel, see *Gyromitra brunnea*
brush-tailed bettong, see *Bettongia penicillata*
Bryoria, 136
Buellia frigida, 143; Plate 70
Bufo periglenes, 30
Burdsall, Harold, 118
butt rot, see *Coniophora puteana*

cabbage, 28
cactophilic, 195
caddisfly, 43
Caesar's mushroom, see *Amanita caesarea*
calcium oxalate, 49
California red-backed vole, see *Myodes californicus*
California University of Pennsylvania, 61
Calvatia, 220
  gigantea, 120, 211

Calypogeiaceae, 177
Calypte anna, Plate 77
Cambrian, 229
Camillea, 97
Cannabis sativa, 32, 33
canopy soil, 159
Cantharellaceae, 118
Cantharellus, 118
   cibarius, 118
   lateritius, 118; Plate 60
capillitium, 149, 150
Capnodiales, 70, 71, 82
carbon balls, see *Daldinia concentrica*
Carboniferous, 229, 235
Carex, 113
Caribbean, 23
Carnegia gigantea, 195
Carroll, Lewis, 210, 211
Caryophyllaceae, 183
Castanea crenata, 213
Castanea dentata, 40, 74, 213, 214
Castanea mollissima, 213
cattle, 28, 49, 51, 169
cedar apple rust, see *Gymnosporangium juniperi-virginianae*
cellophane, 29
cellulase, 167
cellulose, 27, 84, 156, 164–167, 187, 188
Cenozoic, 228
Central America, 23
Cephaloziaceae, 177
Cephaloziellaceae, 177
Ceratiomyxa, 160
   sphaerosperma, Plate 88
Ceratiomyxales, 151
Chaetomium, 84, 167
chanterelle, 22, 115, 118, 119
cherry, 82
chestnut bark disease, see *Cryphonectria parasitica*
chestnut blight, see *Cryphonectria parasitica*

chicken of the woods, see *Laetiporus sulphureus*
China, 23, 110, 126, 213, 218, 219
Chinese chestnut, see *Castanea mollissima*
Ching-Shih, Twang, 147
chipmunk, 214
Chironomidae, Plate 13
chitin, 16, 27
chlamydospore, 40
Chlorociboria, 76
   aeruginascens, Plate 29
chlorophyll, 13, 175, 176
Chlorophyllum, 19, 20
   molybdites, 109
Christopher, John, 182
chytrid, 27–33, 40, 230, 231; Plates 6–8
chytridiomycosis, 30
Chytridiomycota, 20, 22, 27–31, 230
chytridion, 28
Ciboria, 99
   carunculoides, 99
   shiraiana, 99
Cladonia, 135, 137, 140; Plates 76, 78
   convoluta, 131
clamp connection, 102
class, 19
Claviceps, 72
   purpurea, 72
Clavicipitaceae, 72
Clavicipitales, 70, 71
cleistothecium, 69, 70, 75; Plate 28
Clitocybe, 110
   odora, 110
club moss, 38
Coccomyxa, 133
coelomycetes, 56
coenocytic, 15, 47, 55
Collybia, 168
Columbia University, 159

commensalistic relationship, 42
common brown cup, see *Peziza phyllogena*
common earthball, see *Scleroderma citrinum*
common house fly, see *Musca domestica*
common laccaria, see *Laccaria laccata*
conidiogenous cell, 55
conidiophore, 56, 58, 74, 83, 96, 191; Plates 23, 24
conidium, 58, 67, 96
　development of, 55–56
　in spread of diseases, 72, 75, 78, 80, 83, 191
conifer, 165, 167–168, 236
　diseases of, 38, 81, 166
　fossils of, 229, 2330
　fungi and, 114, 126
　lichens and, 142
　mycorrhizal fungi and, 109, 113, 177
　wood wasps and, 192
conifer-cone baeospora, see *Baeospora myosura*
Coniophora puteana, Plate 59
Coprinaceae, 112
Coprinellus, 112
Coprinites dominicana, 236
Coprinopsis, 112
　atramentaria, see *Coprinus atramentarius*
Coprinus, 105, 112, 169, 170, 217
　atramentarius, 112
　comatus, 112; Plate 54
coprophilous, 83, 94, 169, 170
coral fungus, 115, 119, 120; Plate 62
Corallorhiza maculata, 176
Cordyceps, 72, 73
　militaris, 73
　ophioglossoides, 73
　robertsii, 73; Plate 26

sinensis, 73
Coriolus versicolor, 117
cork oak, 40
corn, 28, 59, 181, 184, 225; Plate 103
corn smut fungus, see *Ustilago maydis*
Cornus, 99
　florida, 74
cortex, 132–134, 136, 139, 150, 171, 174, 175, 232
Corticiaceae, 115, 117
corticioid fungus, 115, 117, 118
cortina, 113
Cortinariaceae, 113
Cortinarius, 113, 171
Costa Rica, 30, 57, 154, 188
crape myrtle, 71
Cretaceous, 229, 236, 237
Cribraria cancellata, Plate 85
cross wall, 15, 34, 35
crustacean, 41
crustose lichen, 133, 134, 139, 142, 143; Plate 73
cryophilic, 22
Cryphonectria parasitica, 40, 74, 213
cryptobiosis, 143
Cryptosphaeria, 98
Cryptovalsaria, 99
　americana, 99
　rossica, 99
cuitlacoche, 184
cultivated mushroom, see *Agaricus bisporus*
cup fungus, 69, 92
cyanobacterium, 128, 130, 140, 144, 148
Cyathus striatus, Plate 66
cycad, 38, 229
cyst, 34, 35, 42
cystidium, 105
Cystoderma amianthinum, 168; Plate 93

Cyttaria, 73
  espinosae, 73
  gunnii, 73; Plate 27
Cyttariales, 70, 71, 73

Daldinia, 96, 97
  concentrica, 97, 166; Plate 45
Dawson, Charles, 233, 234
de Bary, Heinrich Anton, 128, 181
De Materia Medica, 221
dead man's fingers, see *Xylaria polymorpha*
Death of Grass, The, 182
deceptive milky, see *Lactarius deliciosus*
deer mushroom, see *Pluteus cervinus*
deer truffle, see *Elaphomyces granulatus*
Dendrobates azureus, 30
Dermatocarpon, 134
  miniatum, Plate 75
desert truffle, see *Terfezia*
destroying angel, see *Amanita bisporigera*
deuteromycetes, 53
Devonian, 229, 230, 233, 234
Diaporthales, 70, 74
Diaporthe, 74
diatom, 23
Diatrypaceae, 96–99
Diatrype, 98
  decorticata, 98
  disciformis, 98
  stigmaoides, 98
  undulata, 98
  virescens, 98; Plate 46
Diatrypella, 98
  placenta, 98
  pulvinata, 98
  quercina, 98
Dictyocaulus, 51
Dictyophora, 124
  indusiata, 218
  multicolor, Plate 67
dictyostelid, 146, 155–162
Dictyostelium, 158
  caveatum, 158
  discoideum, 158
  rosarium, 158
  sphaerocephalum, 158; Plate 90
Diderma, 152
Didymium, 152
Dikarya, 67
dikaryon (dikaryotic), 68, 101, 102, 180, 181
Dionaea muscipula, 196
Dioscorides, Pedanius, 90, 221
diploid condition, 17, 27, 32, 35, 36, 62, 68, 101, 148, 180
discomycetes, 70, 71, 75, 76
Discula destructiva, 74
disease, 14, 219
  caused by ascomycetes, 59, 74, 76, 80–83, 98, 99, 191, 193, 195
  caused by basidiomycetes, 178, 182–185
  caused by chytrids, 29–31
  caused by water molds, 36–37, 39–40
  caused by zygomycetes, 52
Dodgson, Charles, 211
dog, 89, 222, 223
dogwood, see *Cornus*
dong chong zia cao, see *Cordyceps sinensis*
Douglas fir, see *Pseudotsuga menziesii*
downy mildew of grapes, see *Plasmopara viticola*
Drosera, 196
Drosophila, 195
  melanogaster, 195
dry rot fungus, see *Serpula lacrymans*
dryland fish, 86
Dutch elm disease, see *Ophiostoma ulmi*

Dutch still life painters, 209
dye-makers false puffball, see *Pisolithus tinctorius*

earthball, 121, 216
earthstar, 120–123
Eccrinales, 40
Echinodontium tinctorium, 215
Echinosteliales, 150
Echinostelium minutum, Plate 86
Egypt, 220, 225
Elaphomyces, 73, 92
  granulatus, Plate 39
elater, 150
Eleutherodactylus, 113
elk, 51
embryo, 155
Emericella, 60
endobiotic, 29, 230
Endothia parasitica, see *Cryphonectria parasitica*
enokitake, see *Flammulina velutipes*
Enteridium lycoperdon, 153
Entoloma, 111
Entolomataceae, 111
Entomophthora, 51
  maimaiga, 53
  muscae, 51–53; Plate 21
  syrphi, 53
Entomophtherales, 51
epibiotic, 29, 230
epidermis, 69, 74, 172, 231
Equisetum, 157
ergot, 72
Ericaceae, 174, 177, 184
ericoid mycorrhizae, 174–177
Erysiphales, 70, 71, 74
eucalypt, 14, 171, 203, 205
Eucalyptus marginata, 40
Eupenicillium, 60
Europe, 23, 24, 135, 209, 219, 221, 223
  Dutch elm disease in, 81
  edible fungi of, 88–91, 114
  fungi diseases in, 37, 39, 72, 76, 184
  native fungi of, 98, 108, 151
  uses of fungi in 215, 225, 226
European red squirrel, 204
European Space Agency, 129
Eurotiales, 60, 70, 71, 75, 92
Eurotium, 60
Eutypa, 98
Exobasidium, 184; Plate 104
exoenzymes, 16, 27
eyelash cup, see *Scutellinia scutellata*

Fagus grandifolia, 78
fairy ring, 140, 208–210; Plate 113
Fairy Ring, A, 208; Plate 112
fairy ring mushroom, see *Marasmius oreades*
false morel, 87
fermentation, 63, 155, 194, 225, 226
fern, 144, 179, 229
fertilization tube, 35
fir, 171
First Men in the Moon, The, 212
flagellum, 28, 33, 34, 40, 147, 160
Flammulina velutipes, 217
Fleming, Alexander, 220
flower fly, 53
flowering dogwood, see *Cornus florida*
fly agaric, see *Amanita muscaria*
flying squirrel, see *Glaucomys sabrinus*
foliose lichen, 133, 134, 136, 139, 141
Fomes fomentarius, 207, 220
forest fire, 87, 88
fossil, 120, 137, 145, 211, 227–238; Plates 118, 123, and 124
foxfire, 207, 208
Frank, Albert, 172
Fraxinus americana, 86

frog, 30, 31
fruit fly, see *Drosophila*
fruiting bodies, 18, 22, 120, 215, 200, 220, 223
  ascomycetes and, 45, 66–70, 73, 75, 77, 78, 82, 174
  basidiomycetes and, 103–109, 114–126, 190
  cup fungi and, 92–95
  flask fungi and, 95–97
  fossils of, 236, 237
  lichens and, 133
  morels and truffles and, 85–89, 91–92
  slime molds and, 149–150, 153, 154, 157, 160
fruticose lichen, 134, 135, 139, 142
Fuligo septica, 147, 153, 154, 155; Plate 83
Fuligo violacea, 155
fungal biodiversity, 24
fungus-cultivating termite, 189
fungus garden, 187, 188, 191, 192
funiculus, 122, 123
Fusarium, 60

Galerina, 114, 168
gametothallus, 32
Ganoderma applanatum, 116
Ganoderma lucidum, 219
gardenia, 71
gasteromycetes, 119, 120
gastronomic treasure, 88
Geastrum, 121
  saccatum, Plate 65
gem-studded puffball, see *Lycoperdon perlatum*
Genea, 94
genus, 19
Geoglossum, 77
Geopora, 94
germ-pore, 107
giant puffball, see *Calvatia gigantea*

Gilbert's potoroo, see *Potorous gilbertii*
gilled fungus, 104
ginseng, see *Panax*
Glaucomys sabrinus, 203
Gliocladium, 60
Glomeromycota, 20, 230, 232
Glomus, 232
golden chanterelle, see *Cantharellus cibarius*
golden toad, see *Bufo periglenes*
Gorilla berengei, 217
Gorillas in the Mist, 217
grape, 36–38, 64, 76, 226
Graphium, 80
Gray, Asa, 99
gray ghost, 213
gray mold, 76
Great Britain, 24
Great Smoky Mountains National Park, 158
Greeks, 90, 217, 221
green death cap, see *Amanita phalloides*
green stain of wood, see *Chlorociboria aeruginascens*
Greenland, 237, 238
green-spored lepiota, see *Chlorophyllum molybdites*
grouse, 214
Guatemala, 158
gymnosperm, 179, 229
Gymnosporangium juniperi-virginianae, 182; Plate 101
gypsy, see *Rozites caperatus*
gypsy moth, see *Lymantria dispar*
Gyromitra, 87
  brunnea, 87; Plate 37
  esculenta, 88
  gigas, 88
  infula, 88
gyromitrin, 87

hair fungus, 41
Hale, Mason, 133
half-free morel, see *Morchella semilibera*
Hall, Ian, 92
hallucination, 72
hapteron, 123
Harpellales, 41
Hartig, Theodor, 172
Harvard University, 79
hat-thrower, see *Pilobolus*
haustorium, 74, 79, 131
hawthorn, 82
hazelnut, 88, 91
Hebeloma, 114
Helianthemum, 90
Helicoön, Plate 15
Helotiales, 70, 71, 75–77
Helvella, 87
Helvellaceae, 85, 87
Hemitrichia serpula, 149; Plate 87
hemp, see *Cannabis sativa*
Hennebert, Grégoire, 54
Hericium americanum, 119
Hericium erinaceus, 119; Plate 61
hickory, 14, 168
Hohenbuehelia, 199
holdfast, 41, 134
holomorph, 54, 55
honey mushroom, see *Armillaria ostoyae*
honeydew, 71
Horneophyton lignieri, 229
horse, 49, 51, 84, 169, 222
horsetail, see *Equisetum*
hover fly, 53
Hueber, Francis, 234
hummingbird, 136
humongous fungus, 16
Hunter, Barry, 61
Hydnaceae, 115
Hydnangiaceae, 110
Hydnum repandum, 118

Hygrocybe, 110
 cuspidata, Plate 50
Hygrophoraceae, 107, 109, 110
Hygrophorus, 110
hymenium, 68, 69, 77, 86, 89, 93–95, 103–105, 115, 117, 119, 120, 121, 126
Hymenoscyphus, 177
 ericae, 174
hypha, 14–16, 18, 21, 168, 191
 ascomycetes and, 22, 43, 55–57, 66–68, 74, 80, 84
 basidiomycetes and, 22, 101, 102, 105, 114, 116, 120, 122–124
 blastocladiomycetes and, 32, 34, 35, 38
 chytrids and, 27
 decomposing wood and, 165
 fossil evidence of, 228, 231, 234, 237
 infecting myxomycetes and, 154
 insect food and, 44, 193
 lichens and, 131–134, 139
 mycorrhizal associations and, 171–176
 nematodes and, 196–199
 wheat rust and, 181
 zygomycetes and, 40, 41, 47, 48, 52
hyphal tip, 35
hyphal traps, 178
hyphomycetes, 56
Hypocreales, 70, 71, 77, 78, 188
hypogeous fungus, 88, 92, 94, 201, 204
Hypomyces lactifluorum, 77; Plate 31
Hypoxylon, 96, 97
 fragiforme, 97; Plate 44

Illustrated Genera of Imperfect Fungi, 61
Illustrated Guide to the Coprophilous Ascomycetes of Australia, An, 170

imperfect fungus, 53
India, 154, 217
Indian paint fungus, see *Echinodontium tinctorium*
Indian pipe, see *Monotropa uniflora*
Ingold, Cecil Terence, 42
Ingoldian hyphomycetes, 42
Inocybe, 114
inoperculate discomycete, 76
insect, 14, 16, 29, 71–73, 136, 177, 196
    beetles, 153, 190–192, 193–194
    flies, 89, 124, 195–196, 199–200
    fungal spore dispersal and, 153, 159, 183
    leaf-cutter ants, 187
    parasites of, 33, 41–44, 51, 53, 64 78, 79
    wood wasps, 192–193
insect destroyer, see *Entomophthora*
insectivorous plants, 196
iodine, 96, 97
Iran, 225
Ireland, 24, 37
isidium, 139
Italian white truffle, see *Tuber magnatum*

jack-o-lantern mushroom, see *Omphalotus illudens*
Japanese chestnut, see *Castanea crenata*
jarrah, see *Eucalyptus marginata*
jelly club, see *Leotia lubrica*
jelly fungus, 125, 126
jelly tooth, see *Pseudohydum gelantinosum*
Journey to the Center of the Earth, A, 210
Juniperus virginiana, 182
Jurassic, 229
Jurassic Park, 236

kawaratake, see *Trametes versicolor*
Kidston, Robert, 229
king bolete, see *Boletus edulis*
kingdom, 19
kwei hi, 147

Laboulbeniales, 70, 78, 79; Plate 32
Laccaria, 110
    laccata, 110
Lactarius, 77, 109, 171, 175
    deliciosus, 217
Laetiporus sulphureus, 116, 166; Plate 57
Lamproderma, 151
Lang, William, 229
Lange, Jakob, 19
larch, 221
Laricifomes officinalis, 221
late blight of potato, see *Phytophthora infestans*
Laursen, Gary, 168, 204
lawn mower's mushroom, see *Panaeolus foenisecii*
leaf-cutter ant, 187, 188; Plate 105
leather lichen, see *Dermatocarpon miniatum*
Leccinum, 114
    insigne, 114; Plate 56
Leiodidae, 153
Lemonniera centrosphaera, Plate 14
Lentinula edodes, 217, 219
Leotia lubrica, 76; Plate 30
Leotia viscosa, 77
Leotiales, 75, 76
Lepidoderma, 151
Lepidoziaceae, 177
Lepiota, 19, 20, 108
Lepiotaceae, 107–109, 187
Lepista, 110
    nuda, 110
Leptoglossum retirugum, see *Arrhenia retiruga*
Leptonia, 111

Leucoagaricus, 19
Levetin, Estelle, 23
Liceales, 150
lichen, 128, 129, 145, 235; Plates 70–82
   air pollution and, 137–138
   as a dye, 215, 216
   as a food source, 135–137, 202–204
   as an epiphyll, Plate 81
   components of, 21, 130–132, 186
   growth forms of, 132–135
   growth rate, 142–143
   reproduction in, 139–141
   role in succession, 143–144
   substrates for, 141–142
   with cyanobacteria as photobionts, 144–145
lichen agaric, see *Lichenomphalia umbellifera*
lichenometry, 143
Lichenomphalia umbellifera, 133; Plate 72
Lichens of North America, 145
lignin, 164, 165, 187
lilac, see *Syringa vulgaris*
Lindbergh, Charles, 23
ling zhi, see *Ganoderma lucidum*
Liriodendron tulipifera, 86
Lobaria, 133; Plate 74
   oregano, 145
lobster mushroom fungus, see *Hypomyces lactifluorum*
loculoascomycetes, 70, 71, 82
Logan, William, 233
Longcore, Joyce, 30, 31
long-footed potoroo, see *Potorous longipes*
long-nosed potoroo, see *Potorous tridactylus*
Lophoziaceae, 177
luciferase, 207
luciferin, 207

luminescent panellus, see *Panellus stipticus*
Lycogala, 152
   epidendrum, 147
Lycoperdon, 121
   giganteum, 211
   perlatum, 121; Plate 63
   pyriforme, 121
Lymantria dispar, 53

Macaca mulatta, 217
Macquarie Island, 168
macrocyst, 157
macrofungus, 21, 24, 44, 104, 236
Macrofungi Associated with Oaks of Eastern North America, 216
Macrolepiota, 109
   procera, 109
Macrotermes, 189
magic mushroom, see *Panaeolus cubensis*
magnolia, 168
magnolia-cone mushroom, see *Strobilurus conigenoides*
Magnus, Olaus, 208
Maine, 31, 214
maize mushroom, see *Ustilago maydis*
Malheur National Forest, 15
maple, 14
Marasmiaceae, 110, 111
Marasmius, 111
   oreades, 210
   rotula, 111
   siccus, 111; Plate 51
marijuana, see *Cannabis sativa*
Martin, George, 123
Masonhalea richardsonii, 141; Plate 80
mating type, 48, 62, 157, 180, 181
matsutake, see *Tricholoma matsutake*
mayapple, see *Podophyllum peltatum*
mayapple rust, see *Puccinia podophylli*
mazaedium, 140

mead, 226
meadow mushroom, see *Agaricus campestris*
meiosporangium, 32
Merck Research Laboratories, 57
merkel, 86
Merkel, Herman W., 213
Mesopotamia, 223, 225
Mesozoic, 228, 229
Metatrichia, 154
   vesparia, Plate 89
Microbotryum dianthorum, 183; Plate 102
microcyst, 148
Microglossum rufum, 77
Microsphaera penicillata, Plate 28
Microtermes, 189
midge, see Chironomidae
Midsummer Night's Dream, A, 209
migratory birds, 159
Millardet, Alexis, 37
Miller, Orson, 61, 106, 113
millipede, 41, 78
mitosporangium, 32
mitospore, 55
Mitrula paludosa, 44, 45; Plate 16
moist chamber culture, 152, 153
molly moocher, 86
Monacrosporium, 192
Monilia, 76
monokaryotic, 68, 102
monomethylhydrazine, 87
Monotropa uniflora, 174–176; Plate 98
Monotropaceae, 174
Monteverde Cloud Forest Reserve, 30
Monteverde harlequin frog, see *Atelopus varius*
moose, 136
morchel, 86
Morchella, 86
   deliciosa, 86

esculenta, 86; Plate 35
elata, 86; Plate 36
semilibera, 86
Morchellaceae, 85
morel, 22, 85–88, 104, 217
Morgan, Walter Jenks, 209
moth, 73, 136
mound-building termite, 187; Plate 106
mountain goat, 136
mountain gorilla, see *Gorilla berengei*
Mountain Lake Biological Station, 116, 183
mouse, 89
Mucor, 48, 64
mulberry, 99
mule deer, 136
Musca domestica, 52
mushroom, 19, 20, 22, 86, 204, 205, 208–213, 216, 218, 221, 222, 235
mushroom-derived dye, 215, 216; Plate 114
muskox, 136
Mutinus, 124
mutualism, 131, 132, 186
mycelium, 15, 18, 22, 175, 176, 207, 234, 235; Plate 1
   agarics and, 102, 105, 123, 124, 126
   ascomycetes and, 42–44, 55, 56, 58, 62, 67, 71, 74–76, 78, 83, 213
   fairy rings and, 209, 210
   flask fungi and, 96
   nematode-trapping fungi and, 197
   rusts and smuts and, 180–183
   shiitake cultivation and, 218
   slime molds and, 140, 141, 154
   termite fungus and, 188–190
   truffles and, 205
   water molds and, 28, 33, 36–40

wood-decaying fungi and, 193
  zygomycetes and, 47–49, 52
mycelial network, 175
Mycena, 111; Plate 19
  haematopus, 111; Plate 52
  lucentipes, Plate 111
  luteopallens, 168
Mycenaceae, 110, 111
mycobiont, 130–132, 138–141; Plate 71
mycoheterotroph, 175, 176
mycologist, 13, 60, 63, 169
  areas of study, 18–21, 27, 28, 41, 54, 79, 104, 107, 146
  taxonomy and, 71, 95, 107, 110, 150
mycology, 13, 21, 54, 61, 63, 107, 170, 230
mycophagy, 201, 213
mycorrhiza, 170–177; Plates 94–97
mycorrhizal association, 14, 18, 24, 170–177, 186
  of agarics, 108–111, 113
  of *Amanita*, 108
  of ascomycetes, 174, 177
  of basidiomycetes, 171
  of boletes, 114, 115
  of bryophytes, 176–177
  of coral fungi, 120
  of corticioid fungi, 117
  of earthballs, 121
  of honey mushroom, 16
  of orchids, 125, 176
  of polypores, 117
  of truffles, 88, 90, 201, 203, 205
mycotoxin, 59, 60
Myodes californicus, 201
myxomycete, 146–155, 159–162, 167, 237
myxomyceticolous fungus, 153, 154; Plate 89
myxomyceticolous mammal, 200–205; Plate 110

nameko, see *Pholiota nameko*
National Museum of Natural History, 234
National Science Foundation, 20
National Zoological Park, 30
Nectria, 78
  cinnabarina, 78
  coccinea, 78
Nectriopsis violacea, 154
nematode, 51, 158, 178, 196–199, 237; Plate 109
nematode-trapping fungus, 196–198, 237; Plate 109
nematophagous fungus, 196
Neocallimastigomycota, 28
Neurospora, 83, 84
  crassa, 84
New York, 159, 213
New York Botanical Garden, 154
New York Zoological Park, 213
New Zealand, 23, 71, 73, 92, 108, 142, 154, 214
nikau palm, 142
No Blade of Grass, see *The Death of Grass*
noble rot, see *Botrytis cinerea*
Noctuidae, 136
Nolanea, 111
North America, 174, 176, 207
  chestnut blight in, 74, 213–214
  Dutch elm disease in, 81
  edible fungi of, 111, 116
  fungal diseases in, 31, 36, 37, 74, 78, 82, 99, 179, 182, 184
  fungus feeders in, 201, 203
  lichens in, 135, 137, 141
  native fungi of, 23, 24, 77, 86, 92, 93, 98, 151, 158, 159
  poisonous fungi of, 108, 127
northern bettong, see *Bettongia tropica*
northern flying squirrel, see *Glaucomys sabrinus*

northern leopard frog, see *Rana pipiens*
Nostoc, 140, 144
Nothofagus, 71, 73, 171
   menziesii, 145; Plate 27

oak, 14, 40, 88, 91, 98, 108, 171, 208, 218
ochratoxin A, 60
ohia lehua, 40
Olive, Lindsey, 159
Olpidium, 230
Omphalina ericetorum, see *Lichenomphalia umbellifera*
Omphalotus illudens, 207
oogonium, 35, 36, 40; Plate 12
oomycete, 41
Oomycota, 21
oosphere, 35, 36
oospore, 36, 38; Plate 12
operculate discomycete, 76
operculum, 29, 76
Ophiostoma, 79–81
   novo-ulmi, 81
   piceae, 81
   ulmi, 80, 81
Ophiostomatales, 70, 79
orange club, see *Cordyceps militaris*
orange earth tongue, see *Microglossum rufum*
orange pinwheel marasmius, see *Marasmius siccus*
orchid, 125, 176
Orchidaceae, 176
order, 19
Ordovician, 229, 234
owlet moth, see Noctuidae
oyster mushroom, see *Pleurotus*

Pacific Ocean, 23
Palaeoblastocladis milleri, 231; Plate 119
Paleopyrenomycites devonicus, 231

Paleozoic, 228, 229
palm, 24, 142, 167
Panaeolus foenisecii, 222; Plate 116
Panama, 30, 194
Panax, 99
Panckow, Thomas, 147
Panellus stipticus, 207, 208
parasexual reproduction, 17
parasite, 28, 33, 36, 42, 131, 154, 164, 196, 197
   defined, 14
   obligate, 56, 74, 179, 183
   of animals, 47, 57
   of fungus, 71, 77, 126, 188
   of humans, 47, 57
   of insects, 51, 71, 73
   of plants, 47, 57, 77, 103, 125, 178, 184, 231
   of woody plants, 78, 97, 99, 110
Parasola, 112
Parmelia, 132, 133, 139
   omphalodes, 215
   saxatilis, 215
partial veil, 106, 113
Pasteur, Louis, 63
pathogen, 14, 47, 60
   of agricultural crops, 76, 216
   of animals, 30, 32, 47, 60
   of humans, 32, 47, 60
   of plants, 32, 38, 71, 72, 95, 103, 178, 183
   of trees, 39, 74, 78, 80, 97, 98, 99, 213
Paxillaceae, 113
Paxillus, 113
pear-shaped puffball, see *Lycoperdon perlatum*
Peltigera, 139
penicillin, 59, 220
Penicillium, 58–69, 75, 220; Plate 24
   camemberti, 59
   chrysogenum, 59
   digitatum, Plate 17

roqueforti, 59
peridium, 89, 91, 92, 120–122, 124, 149, 150; Plate 64
Périgord black truffle, see *Tuber melanosporum*
perithecium, 69, 70, 72, 74, 77, 78, 80, 82, 84, 85, 95, 96, 140, 155, 231; Plate 34
Permian, 229
Peziza, 93
 phyllogena, 93
Pezizaceae, 85, 93
Pezizales, 71, 75, 76, 85, 92, 95, 96
Phallus, 124
phialide, 58; Plate 23
Phialophoropsis, 192
Pholiota nameko, 218
Phomopsis, 74
photobiont, 130–132, 138–141, 144; Plate 71
photosynthesis, 14, 16, 130, 131, 164, 170, 175
Phragmidium mucronatum, Plate 3
Phycomycetes, 26, 27, 40
Phyllactinia, 75
Physalacriaceae, 110
Physarales, 150
Physarum polycephalum, 167
Physcia, 133
Phytophthora cinnamomi, 38–40
Phytophthora infestans, 36–38
Picea glauca, 114
Picea rubens, 144
pig, 89
Pilát, Albert, 19
Pilobolaceae, 49
Pilobolus, 49–51, 169; Plate 20
pin lichen, 140
pine, 14, 29, 40, 81, 88, 114, 171
pine mushroom, see *Tricholoma matsutake*
pink bottom, see *Agaricus campestris*
Pinus, see pine

Pinus longaeva, 164
pinwheel mushroom, see *Marasmius rotula*
Piptoporus betulinus, 165, 207, 220
Pisolithus tinctorius, 216; Plate 115
pixie-cup lichen, 140; Plate 78
plasmodiocarp, 149; Plate 87
plasmodium, 148, 149, 151, 160, 162, 167; Plate 84
Plasmopara viticola, 36–38
plectomycete, 70, 71
Pleosporales, 70, 71, 81
Pleurotus, 199, 218
Pliny the Elder, 90, 208, 225
Pluteaceae, 111
Pluteus cervinus, 111; Plate 53
Podophyllum, 99
 peltatum, 179
Podospora, 84
 fimiseda, Plate 34
Polishook, Jon, 57
pollen grain, 23, 29, 230; Plate 8
Polycephalomyces tomentosus, 154; Plate 89
Polyporaceae, 115, 117
polypore, 22, 104, 115–118, 166, 178, 201, 207, 215, 220, 221
Polysphondylium, 157, 158
 pallidum, 158; Plate 91
polysporous, 98
poplar, 91, 218
Populus tremuloides, 16, 114
portobello mushroom, see *Agaricus blazei*
potato, 28, 36–38, 182
Potorous gilbertii, 203
Potorous longipes, 202; Plate 110
Potorous tridactylus, 203
powdery mildew, 74, 75
Precambrian, 228, 231
Protomycena electra, 236; Plate 123
protostelid, 146, 150, 159–161
Protostelium mycophaga, 159, 161; Plate 92

Prototaxites, 233–235; Plates 121, 122
protozoan, 28, 40, 158, 188
Prunus serotina, 82
Psalliota, 19
pseudoaethalium, 149
Pseudohydnum gelantinosum, 126; Plate 69
pseudomycelium, 62
pseudoplasmodium, 156
pseudorhiza, 190
pseudothecium, 70, 82
Pseudotsuga menziesii, 145
Psilocybe, 112, 221, 222
    cubensis, 222
    mexicana, 221
Puccinia, 179
    graminis, 179, 180; Plate 100
    podophylli, 179; Plate 99
Pucciniomycotina, 102, 103, 178
Puerto Rico, 154
puffball, 22, 120–122, 216, 217, 220
pycnidium, 56, 74
pyrenomycete, 70, 71, 95, 231
Pyronemataceae, 85, 93
Pyrus malus, 15, 82, 86, 182

quail, 214
quaking aspen, see *Populus tremuloides*

raccoon, 214
radioisotopic date, 228
Raffaelea, 192
Rana pipiens, Plate 9
red cedar, see *Juniperus virginiana*
red cushion hypoxylon, see *Hypoxylon fragiforme*
red spruce, see *Picea rubens*
red squirrel, see *Tamiasciurus hudsonicus*
reindeer, 135–137
reindeer lichen, see *Cladonia*

reishi, see *Ganoderma lucidum*
rhesus monkey, see *Macaca mulatta*
rhizine, 134
Rhizocarpon geographicum, 129
Rhizoctonia, 125
rhizoid, 28, 32, 33, 47, 177
Rhizophydium, 230
Rhizopus stolonifer, 47, 48, 62; Plate 18
Rhodosporidium, 63
    toruloides, 63
Rhodotorula, 63, 237
Rhynia gwynne-vaughani, 229, 233
Rhynie chert, 229–233, 235
Roccella, 215
    tinctoria, 215
Rocky Mountains, 151
rodent, 114, 159, 178
Rogerson, Clark, 154
Romans, 90, 181, 217, 221, 225
Rosaceae, 82
rose rust, see *Phragmidium mucronatum*
rotifer, 178
round earthstar, see *Geastrum saccatum*
roundworm, 177
Rozites, 114
    caperatus, 114
rumen, 28
Russula, 77, 109, 171, 175; Plate 49
    alutacea, 217
Russulaceae, 107, 109
Russulales, 107
rye grass, see *Secale cereale*

Saccardo, Pier Andrea, 60, 61
Saccharomyces cerevisiae, 62–64, 224; Plate 25
Saccharomyces pasteurianus, 224
Saccharomycotina, 56
saffron powder-cap, see *Cystoderma amianthinum*

saguaro cactus, see *Carnegia gigantea*
Sancho, Leopoldo, 129
Saprolegnia, 34, 35; Plate 11
Saprolegniales, 34
saprotroph, 14, 164, 168, 175, 178, 196, 210, 218, 222
　ascomycetes and, 42, 56–58, 74, 76–78, 80, 231
　basidiomycetes and, 93, 95–97, 110–112, 114, 117, 120, 121, 125
　blastocladiomycetes and, 31–33
　chytrids and, 27
　myxomycetes and, 154
　yeasts and, 64
　zygomycetes and, 47
Sarcoscypha occidentalis, 93; Plate 40
Sarcosomataceae, 85, 93
scaber, Plate 56
scale insect, 71
Scapaniaceae, 177
scarlet waxy cup, see *Hygrocybe cuspidata*
Schwendener, Simon, 129
Scleroderma citrinum, 121; Plate 64
sclerotium, 72, 148
Scutellinia, 94
　scutellata, 93, 94; Plate 41
Secale cereale, 72
seed fern, 229
Senecio, 113
septum, 15, 27, 35, 40, 41, 48, 49, 55, 79, 125, 234
Serpula lacrymans, 166
sexual reproduction, 17–18, 29, 34–36, 40–41, 47–48, 62, 67–69, 156–157
shaggy mane, see *Coprinus comatus*
Shakespeare, William, 209
Sharnoff, Stephen, 145
Sharnoff, Sylvia, 145
sheep, 28, 49, 214, 222

shield lichen, see *Parmelia*
shiitake, see *Lentinula edodes*
shortleaf pine, 40
shredder, 43, 44
shrew, 89
Silene alba, 183; Plate 102
Silurian, 229
silver ear, see *Tremella fuciformis*
Simuliidae, 42
Singh, Lal, 217
Sirex, 192
slime mold, 21, 146–162
slime mold beetle, 153
slug, 136, 178
Smittium morbosum, 42
Smittium simulii, Plate 13
smooth chanterelle, see *Cantharellus lateritius*
Snowy Mountains, 151
social insects, 187
Soma: Divine Mushroom of Immortality, 223
sooty mold, 71
Sordaria, 84
　fimicola, 84
Sordariales, 70, 83
soredium, 139
sorus, 157
South America, 23, 36, 73, 187, 221
spermogonium, 181
sporocarp, 149, 173
sporophore, 149, 234, 235
southern beech, see *Nothofagus*
Spathularia, 77
spermogonium, 181
Spinellus fusiger, 47; Plate 19
sponge mushroom, 86
sporangiophore, 36, 38, 47, 49–52
spore print, 105, 109, 114
sporothallus, 32
spotted coral root, see *Corallorhiza maculata*
spring agaricus, see *Agaricus bitorquis*

spruce, 92, 171, 177, 204
squamulose lichen, 132, 134, 135
St. Anthony's fire, 72
stalked scarlet cup, see *Sarcoscypha occidentalis*
starfish stinkhorn, see *Aseroë rubra*
Stemonitales, 150
Stemonitis, 152, 237
Stereaceae, 117
Stereum, 126, 193
sterigma, 101, 105
stinkhorn, 120, 123–125
Stoianovitch, Carmen, 159
stonefly, 43
straw mushroom, see *Volvariella volvacea*
Strawberry Mountains, 15, 16
Strobilurus conigenoides, 168
stroma, 73, 77, 95–98
Strophariaceae, 112
stubble lichen, 140
Suillia, 89
Suillus, 114
sulfur shelf, see *Laetiporus sulphureus*
sundew, see *Drosera*
swamp beacon, see *Mitrula paludosa*
sweet tooth, see *Hydnum repandum*
symbiosis, 131
synnema, 80, 154
Syringa vulgaris, Plate 28
syrphid fly, 53

Talaromyces, 60
tall oat grass, see *Arrhenatherum elatius*
Tamiasciurus hudsonicus, 204
tannin, 213, 214
Taphrinomycotina, 56
tardigrade, 178
Tasmania, 92
Tasmanian bettong, see *Bettongia gaimardi*
taxon, 20

Taylor, John, 227
teleomorph, 54, 55, 57, 60, 63, 75, 154
teliospore, 63, 179, 180, 182–184
telium, 180
Tempest, The, 209
teonanacatl, 221
Terfezia, 90
termite fungus, see *Termitomyces*
Termitomyces, 189; Plate 107
  titanicus, 190
tetraradiate spore, 43; Plate 14
thallus, 28, 29, 32, 33, 41, 79, 128, 130–144
Thaxter, Roland, 79
Theophrastus, 90, 217
Tikal, 158
tinsel flagellum, 33
tooth fungus, 115, 118, 119, 126
Trametes versicolor, 117, 166, 219; Plate 58
Trebouxia, 130, 141
tree fern, 167
Tremella, 125
  fuciformis, 126, 218
  mesenterica, 125; Plate 68
Tremellales, 125
Triassic, 229
Trichia, 152, 154
Trichiales, 150, 154
Trichoderma, 60
Trichoglossum, 77
Tricholoma, 110
  matsutake, 110
Tricholomataceae, 107, 110, 112
trichomycete(s), 40–42; 47, Plate 13
trichospore, 41, 42
true fly, 43
truffle, 22, 88–92, 94, 104, 201–205, 217; Plate 38
truffle fly, see *Suillia*
Tuber, 88, 90, 94
  magnatum, 91

melanosporum, 91; Plate 38
Tuberaceae, 85
Tuberales, 88
tulip-poplar, see *Liriodendron tulipifera*
tumbleweed, 141
turkey, 59
turkey tail, see *Trametes versicolor*
turkey X disease, 59

Ulmus americana, 81
Umbilicaria, 134, 137
umbilicus, 134
Uncinula, 75
unilocular fruiting body, 82
United States, 23, 39, 83, 125, 158, 219
universal veil, 106, 108, 109
Universidad Complutense of Madrid, 129
University of Bordeaux, 37
University of Chicago, 234
University of Maine, 30
University of Tulsa, 23
University of Virginia, 183
uredinium, 180
urediospore, 180
Ustilaginomycotina, 102, 103
Ustilago avenae, 184
  maydis, 184; Plate 103
  violacea, 183
Utah, 16

VAM fungus, 173
van Leeuwenhoek, Anton, 63
vascular plants, 24, 164, 165, 187
  fossils of, 230–234
  lichens and, 131, 141, 142
  mycorrhizal associations of, 170, 174–177
  myxomycetes and, 152
  parasites of, 28, 179
Vasilyeva, Larissa, 99

vegetable caterpillar fungus, see *Cordyceps sinensis* and *C. robertsii*
Venturia inaequalis, 82
Venus flytrap, see *Dionaea muscipula*
Verne, Jules, 210, 211
Verticillium, 60
  rexianum, 154
vesicle, 49, 50, 58, 173; Plate 96
Victorian fairy paintings, 209
Virginia, 116, 161, 183, 210, 214
Volvariella volvacea, 217

walnut, 168
walnut mycena, see *Mycena luteopallens*
Washington, D.C., 30
Wasson, Gordon, 223
water mold, 21, 26, 27, 33–36, 38, 40; Plates 10–12
Weresub, Luella, 54
West Virginia, 92
West Virginia University, 61, 144, 158
wheat, 72, 179–182, 225
wheat rust, see *Puccinia graminis*
whiplash flagellum, 33
white campion, see *Silene alba*
white jelly fungus, see *Tremella fuciformis*
white morel, see *Morchella deliciosa*
white spruce, see *Picea glauca*
white truffle, see *Tuber magnatum*
whitefly, 71
whitetail deer, 92, 136
wild cherry, see *Prunus serotina*
wine, 37, 76, 221, 226
witches' butter, see *Tremella mesenterica*
wood blewit, see *Lepista nuda*
wood ear, see *Auricularia auricularjudae*

Xanthoparmelia, 141
Xanthoria elegans, 129
Xenopus, 31
   laevis, 31
Xylaria, 96
   polymorpha, 96; Plate 43
Xylariaceae, 96–98
Xylariales, 71, 85, 95, 96
Xyleborinus saxesenii, Plate 108

yeast, 55, 56, 61–65, 148, 193–195, 224–226, 238
yellow fairy cup, see *Bisporella citrine*
yellow morel, see *Morchella esculenta*
yellow-gilled russula, see *Russula alutacea*
Youd, Samuel, 182
yun-zhi, see *Trametes versicolor*

Zambonelli, Alessandra, 92
zoosporangium, 28, 29, 34, 35, 40, 231; Plates 7, 10, 11
zoospore, 28, 29, 32–35, 37–39; Plate 7
zygomycete(s), 40, 47–49, 51, 196
Zygomycota, 20, 22, 27, 40, 46, 47, 55, 64, 169
zygote, 29, 148, 155